Lecture Notes in Electrical Engineering

Volume 450

About this Series

"Lecture Notes in Electrical Engineering (LNEE)" is a book series which reports the latest research and developments in Electrical Engineering, namely:

- Communication, Networks, and Information Theory
- Computer Engineering
- Signal, Image, Speech and Information Processing
- Circuits and Systems
- Bioengineering

LNEE publishes authored monographs and contributed volumes which present cutting edge research information as well as new perspectives on classical fields, while maintaining Springer's high standards of academic excellence. Also considered for publication are lecture materials, proceedings, and other related materials of exceptionally high quality and interest. The subject matter should be original and timely, reporting the latest research and developments in all areas of electrical engineering.

The audience for the books in LNEE consists of advanced level students, researchers, and industry professionals working at the forefront of their fields. Much like Springer's other Lecture Notes series, LNEE will be distributed through Springer's print and electronic publishing channels.

More information about this series at http://www.springer.com/series/7818

Kuinam J. Kim · Hyuncheol Kim
Nakhoon Baek
Editors

IT Convergence and Security 2017

Volume 2

Editors
Kuinam J. Kim
Kyonggi University
Seongnam-si, Kyonggi-do
Korea (Republic of)

Hyuncheol Kim
Computer Science
Namseoul University
Cheonan, Chungcheongnam-do
Korea (Republic of)

Nakhoon Baek
School of Computer Science
and Engineering
Kyungpook National University
Daegu
Korea (Republic of)

ISSN 1876-1100 ISSN 1876-1119 (electronic)
Lecture Notes in Electrical Engineering
ISBN 978-981-13-4882-2 ISBN 978-981-10-6454-8 (eBook)
DOI 10.1007/978-981-10-6454-8

Softcover re-print of the Hardcover 1st edition 2018

Printed on acid-free paper

This Springer imprint is published by Springer Nature
The registered company is Springer Nature Singapore Pte Ltd.
The registered company address is: 152 Beach Road, #21-01/04 Gateway East, Singapore 189721, Singapore

Preface

This LNEE volume contains the papers presented at the iCatse International Conference on IT Convergence and Security (ICITCS 2017) which was held in Seoul, South Korea, during September 25 to 28, 2017.

The conferences received over 200 paper submissions from various countries. After a rigorous peer-reviewed process, 69 full-length articles were accepted for presentation at the conference. This corresponds to an acceptance rate that was very low and is intended for maintaining the high standards of the conference proceedings.

ICITCS2017 will provide an excellent international conference for sharing knowledge and results in IT Convergence and Security. The aim of the conference is to provide a platform to the researchers and practitioners from both academia and industry to meet the share cutting-edge development in the field.

The primary goal of the conference is to exchange, share and distribute the latest research and theories from our international community. The conference will be held every year to make it an ideal platform for people to share views and experiences in IT Convergence and Security-related fields.

On behalf of the Organizing Committee, we would like to thank Springer for publishing the proceedings of ICITCS2017. We also would like to express our gratitude to the 'Program Committee and Reviewers' for providing extra help in the review process. The quality of a refereed volume depends mainly on the expertise and dedication of the reviewers. We are indebted to the Program Committee members for their guidance and coordination in organizing the review process and to the authors for contributing their research results to the conference.

Our sincere thanks go to the Institute of Creative Advanced Technology, Engineering and Science for designing the conference Web page and also spending countless days in preparing the final program in time for printing. We would also like to thank our organization committee for their hard work in sorting our manuscripts from our authors.

We look forward to seeing all of you next year's conference.

Kuinam J. Kim
Nakhoon Baek
Hyuncheol Kim
Editors of ICITCS2017

Organizing Committee

General Chairs

Hyung Woo Park	KISTI, Republic of Korea
Nikolai Joukov	New York University and modelizeIT Inc, USA
Nakhoon Baek	Kyungpook National University, Republic of Korea
HyeunCheol Kim	NamSeoul University, Republic of Korea

Steering Committee

Nikolai Joukov	New York University and modelizeIT Inc, USA
Borko Furht	Florida Atlantic University, USA
Bezalel Gavish	Southern Methodist University, USA
Kin Fun Li	University of Victoria, Canada
Kuinam J. Kim	Kyonggi University, Republic of Korea
Naruemon Wattanapongsakorn	King Mongkut's University of Technology Thonburi, Thailand
Xiaoxia Huang	University of Science and Technology Beijing, China
Dato' Ahmad Mujahid Ahmad Zaidi	National Defence University of Malaysia, Malaysia

Program Chair

Kuinam J. Kim	Kyonggi University, Republic of Korea

Publicity Chairs

Miroslav Bureš	Czech Technical University, Czech Republic
Dan (Dong-Seong) Kim	University of Canterbury, New Zealand
Sanggyoon Oh	BPU Holdings Corp, Republic of Korea
Xiaoxia Huang	University of Science and Technology Beijing, China

Financial Chair

Donghwi Lee	Dongshin University, Republic of Korea

Publication Chairs

Minki Noh	KISTI, Republic of Korea
Hongseok Jeon	ETRI, Republic of Korea

Organizers and Supporters

Institute of Creative Advanced Technologies, Science and Engineering
Korea Industrial Security Forum
Korean Convergence Security Association
University of Utah, Department of Biomedical Informatics, USA
River Publishers, Netherlands
Czech Technical University, Czech Republic
Chonnam National University, Republic of Korea
University of Science and Technology Beijing, China
King Mongkut's University of Technology Thonburi, Thailand
ETRI, Republic of Korea
KISTI, Republic of Korea
Kyungpook National University, Republic of Korea
Seoul Metropolitan Government

Program Committee

Bhagyashree S R	ATME College of Engineering, Mysore, Karnataka, India
Richard Chbeir	Université Pau & Pays Adour (UPPA), France
Nandan Mishra	Cognizant Technology Solutions, USA
Reza Malekian	University of Pretoria, South Africa
Sharmistha Chatterjee	Florida Atlantic University, USA

Anooj P.K.	Al Musanna College of Technology, Oman
Hyo Jong Lee	Chonbuk National University,Korea
D'Arco Paolo	University of Salerno, Italy
Suresh Subramoniam	CET School of Management, India
Abdolhossein Sarrafzadeh	Unitec Institute of Technology, New Zealand
Stelvio Cimato	University of Milan, Italy
Ivan Mezei	University of Novi Sad, Serbia
Terje Jensen	Telenor, Norway
Selma Regina Martins Oliveira	Federal Fluminense University, Brazil
Firdous Kausar	Imam Ibm Saud University, Saudi Arabia
M. Shamim Kaiser	Jahangirnagar University, Bangladesh
Maria Leonilde Rocha Varela	University of Minho, Portugal
Nadeem Javaid	COMSATS Institute of Information Technology, Pakistan
Urmila Shrawankar	RTM Nagpur University, India
Yongjin Yeom	Kookmin University, Korea
Olivier Blazy	Université de Limoges, France
Bikram Das	NIT Agartala, India
Edelberto Franco Silva	Universidade Federal de Juiz de Fora, Brazil
Wing Kwong	Hofstra University, USA
Dae-Kyoo Kim	Oakland University, USA
Nickolas S. Sapidis	University of Western Macedonia, Greece
Eric J. Addeo	DeVry University, USA
T. Ramayah	Universiti Sains Malaysia, Malaysia
Yiliu Liu	Norwegian University, Norway
Shang-Ming Zhou	Swansea University, UK
Anastasios Doulamis	National Technical University, Greece
Baojun Ma	Beijing University, China
Fatemeh Almasi	Ecole Centrale de Nantes, France
Mohamad Afendee Mohamed	Universiti Sultan Zainal Abidin, Malaysia
Jun Peng	University of Texas, USA
Nestor Michael C. Tiglao	University of the Philippines Diliman, Philippines
Mohd Faizal Abdollah	University Technical Malaysia Melaka, Malaysia
Alessandro Bianchi	University of Bari, Italy
Reza Barkhi	Virginia Tech, USA
Mohammad Osman Tokhi	London South Bank University, UK
Prabhat K. Mahanti	University of New Brunswick, Canada
Chia-Chu Chiang	University of Arkansas at Little Rock, USA
Tan Syh Yuan	Multimedia University, Malaysia
Qiang (Shawn) Cheng	Southern Illinois University, USA
Michal Choras	University of Science and Technology, Korea
El-Sayed M. El-Alfy	King Fahd University, Saudi Arabia
Abdelmajid Khelil	Landshut University, Germany

James Braman	The Community College of Baltimore County, USA
Rajesh Bodade	Defence College of Telecommunication Engineering, India
Nasser-Eddine Rikli	King Saud University, Saudi Arabia
Zeyar Aung	Khalifa University, United Arab Emirates
Schahram Dustdar	TU Wien, Austria
Ya Bin Dang	IBM Research, China
Marco Aiello	University of Groningen, Netherlands
Chau Yuen	Singapore University, Singapore
Yoshinobu Tamura	Tokyo City University, Japan
Nor Asilah Wati Abdul Hamid	Universiti Putra Malaysia, Malaysia
Pavel Loskot	Swansea University, UK
Rika Ampuh Hadiguna	Andalas University, Indonesia
Hui-Ching Hsieh	Hsing Wu University, Taiwan
Javid Taheri	Karlstad University, Sweden
Fu-Chien Kao	Da-Yeh University, Taiwan
Siana Halim	Petra Christian University, Indonesia
Goi Bok Min	Universiti Tunku Abdul Rahman, Malaysia
Shamim H Ripon	East West University, USA
Munir Majdalawieh	George Mason University, USA
Hyunsung Kim	Kyungil University, Korea
Ahmed A. Abdelwahab	Qassim University, Saudi Arabia
Vana Kalogeraki	Athens University, Greece
Joan Ballantine	Ulster University, UK
Jianbin Qiu	Harbin Institute of Technology, China
Mohammed Awadh Ahmed Ben Mubarak	Infrastructure University Kuala Lumpur, Malaysia
Mehmet Celenk	Ohio University, USA
Shakeel Ahmed	King Faisal University, Saudi Arabia
Sherali Zeadally	University of Kentucky, USA
Seung Yeob Nam	Yeungnam University, Korea
Tarig Mohamed Hassan	University of Khartoum, Sudan
Vishwas Ruamurthy	Visvesvaraya Technological University, India
Ankit Chaudhary	Northwest Missouri State University, USA
Mohammad Faiz Liew Abdullah	University Tun Hussein Onn, Malaysia
Francesco Lo Presti	University of Rome Tor Vergata, Italy
Muhammad Usman	National University of Sciences and Technology (NUST), Pakistan
Kurt Kurt Tutschku	Blekinge Institute of Technology, Sweden
Ivan Ganchev	University of Limerick, Ireland/University of Plovdiv “Paisii Hilendarski”
Mohammad M. Banat	Jordan University, Jordan

David Naccache	Ecole normale supérieure, France
Kittisak Jermsittiparsert	Rangsit University, Thailand
Pierluigi Siano	University of Salerno, Italy
Hiroaki Kikuchi	Meiji University, Japan
Ireneusz Czarnowski	Gdynia Maritime University, Poland
Lingfeng Wang	University of Wisconsin-Milwaukee, USA
Somlak Wannarumon Kielarova	Naresuan University, Thailand
Chang Wu Yu	Chung Hua University, Taiwan
Kennedy Njenga	University of Johannesburg, Republic of South Africa
Kok-Seng Wong	Soongsil University, Korea
Ray C.C. Cheung	City University of Hong Kong, China
Stephanie Teufel	University of Fribourg, Switzerland
Nader F. Mir	San Jose State University, California
Zongyang Zhang	Beihang University, China
Alexandar Djordjevich	City University of Hong Kong, China
Chew Sue Ping	National Defense University of Malaysia, Malaysia
Saeed Iqbal Khattak	University of Central Punjab, Pakistan
Chuangyin Dang	City University of Hong Kong, China
Riccardo Martoglia	FIM, University of Modena and Reggio Emilia, Italy
Qin Xin	University of the Faroe Islands, Faroe Islands, Denmark
Andreas Dewald	ERNW Research GmbH, Germany
Rubing Huang	Jiangsu University, China
Sangseo Parko	Korea
Mainguenaud Michel	Insitut National des sciences Appliquées Rouen, France
Selma Regina Martins Oliveira	Universidade Federal Fluminense, Brazil
Enrique Romero-Cadaval	University of Extremadura, Spain
Noraini Che Pa	Universiti Putra Malaysia (UPM), Malaysia
Minghai Jiao	Northeastern University, USA
Ruay-Shiung Chang	National Taipei University of Business, Taiwan
Afizan Azman	Multimedia University, Malaysia
Yusmadi Yah Jusoh	Universiti Putra Malaysia, Malaysia
Daniel B.-W. Chen	Monash University, Australia
Wuxu Peng	Texas State University, USA
Noridayu Manshor	Universiti Putra Malaysia, Malaysia
Alberto Núñez Covarrubias	Universidad Complutense de Madrid, Spain
Stephen Flowerday	University of Fort Hare, Republic of South Africa
Anton Setzer	Swansea University, UK
Jinlei Jiang	Tsinghua University, China

Lorna Uden	Staffordshire University, UK
Wei-Ming Lin	University of Texas at San Antonio, USA
Lutfiye Durak-Ata	Istanbul Technical University, Turkey
Srinivas Sethi	IGIT Sarang, India
Edward Chlebus	Illinois Institute of Technology, USA
Siti Rahayu Selamat	Universiti Teknikal Malaysia Melaka, Malaysia
Nur Izura Udzir	Universiti Putra Malaysia, Malaysia
Jinhong Kim	Seoil University, Korea
Michel Toulouse	Vietnamese-German University, Vietnam
Vicente Traver Salcedo	Universitat Politècnica de València, Spain
Hardeep Singh	Ferozepur College of Engg & Technology (FCET) India, India
Jiqiang Lu	Institute for Infocomm Research, Singapore
Juntae Kim	Dongguk University, Korea
Kuo-Hui Yeh	National Dong Hwa University, China
Ljiljana Trajkovic	Simon Fraser University, Canada
Kouichi SAKURAI	Kyushu Univ., Japan
Jay Kishigami	Muroran Institute of Technology, Japan
Dachuan Huang	Snap Inc., USA
Ankit Mundra	Department of IT, School of Computing and IT, Manipal University Jaipur, India
Hanumanthappa J.	University of Mysore, India
Muhammad Zafrul Hasan	Texas A&M International University, USA
Christian Prehofer	An-Institut der Technischen Universitaet Muenchen, Germany
Lim Tong Ming	Sunway University, Malaysia
Yuhuan Du	Software Engineer, Dropbox, San Francisco, USA
Subrata Acharya	Towson University, USA
Warusia Yassin	Universiti Teknikal Malaysia Melaka, Malaysia
Fevzi Belli	Univ. Paderborn, Germany

Contents

Security and Privacy

Intelligent Vehicular Networking and Applications

Investigating Public Opinion Regarding Autonomous Vehicles: A Perspective from Chiang Mai, Thailand

Kenneth Cosh(✉), Sean Wordingham, and Sakgasit Ramingwong

Computer Engineering Department, Faculty of Engineering,
Chiang Mai University, Chiang Mai, Thailand
drkencosh@gmail.com

Abstract. Autonomous vehicles (AVs) are fast becoming a realistic vision of the near future. With many technological advances being made, AVs are already in the testing phase in several locations around the developed world. This research project investigates the prospect of AVs in a developing country, specifically in the city of Chiang Mai, Thailand, where driving conditions are significantly different from those in the locations where the technology is currently being tested. A broad section of the public was surveyed to investigate awareness and opinions concerning AVs in general and concerning their use within Chiang Mai. The survey presented here finds a city interested in the potential of the technology, particularly the prospect of better safety.

Keywords: Autonomous Vehicles · Thailand

1 Introduction

Autonomous vehicles (AVs) have for a long time been a feature of science fiction, but recent technological advances and investment in research are bringing into the near future the possibility of driverless travel. Much effort has been made to solve the technical challenges of what is likely to become the latest massively disruptive technological advance. Self-driving cars are already being tested in several countries, mainly the United States and Singapore, with developments being driven by leading technology companies such as Google and virtually all car manufacturers, notably Tesla. This research investigates the future of autonomous vehicles in a developing country, Thailand.

There are many advantages being put forward by the proponents of AVs, including improved road safety, reduction in the cost of travel [1], new opportunities for the disabled or elderly [2], and also the potential for minors to travel without adult supervision. With fewer vehicles needing to be on the road journeys may be faster, and AVs may have a positive impact on the environment [3]. Allowing AVs to communicate with each other and plan their most efficient routes, congestion will be reduced. As vehicles will no longer need to be parked close to their owners, cities can be redesigned and space reallocated [4]. Indeed AVs may allow alternative choices with people choosing to live further away from their workplace and make use of a productive commute [5].

K.J. Kim et al. (eds.), *IT Convergence and Security 2017*,
Lecture Notes in Electrical Engineering 450,
DOI 10.1007/978-981-10-6454-8_1

AVs are likely to be an incredibly disruptive technology, causing a radical shift in the way the world operates. As well as the likely impacts on the layout and design of the cities of the future, and the social impacts of changing lifestyles, certain industries and careers are likely to be impacted. Clearly many roles within transportation industries will be affected, including public transport and logistics. Cost savings through logistics efficiencies will have knock-on effects to the cost of products. Meanwhile industries such as insurance are likely to be affected through reductions in accidents and claims.

The exact nature of future driving is unknown, with usage likely to evolve through a variety of models. There are several stages of automation, from function-specific automation, where particular functions such as cruise control and automated parking are available, to level 2 automation where multiple functions are combined, allowing adaptive cruise control and lane centering, but where the driver is expected to be in control at all times. Level 3 allows limited self-driving automation, where all safety critical functions are under the control of the automobile, up to full self-driving automation, where occupants are not able to take over the controls of the vehicle [6].

Concerning personal transportation, there are further alternative visions. Some may prefer to own their personal autonomous vehicle, while in some places the increasing popularity of ride sharing might make mobility on-demand, pay-per-use services more popular, when offered at an affordable price [7]. The eventual usage model may ultimately depend on government policy, legal decisions and the preferences of the general public. Various opinion surveys have been conducted to assess public opinion, but these surveys have focused on the opinions of developed western countries, mainly the United States, United Kingdom and Australia [8–10] with few taking a global perspective [11] and some focus on the importance of trust regarding AVs [12]. This paper focuses on the city of Chiang Mai, in northern Thailand, a city with a different transportation infrastructure from most western cities.

Driving conditions in Thailand are very different from those elsewhere in the world. A World Health Organisation report in 2015 ranked Thailand's roads as the second deadliest in the world, with 36.2 fatalities per 100,000 inhabitants per year [13]. Road traffic injuries are the leading cause of death amongst adolescents and young adults [14] and road traffic accidents the second leading cause of death overall [15]. Most of these accidents (74%) involved a motorcycle and there is a significant association with speeding, not wearing a helmet, and alcohol consumption despite a Compulsory Helmet Law passed in 1994–1995 [16, 17]. Between 2004 and 2012 the number of registered motor vehicles rose from 19.8 to 31.4 million, and during this period the ratio of cars to motorbikes increased, with car driving licenses increasing by 41% and motorcycle license declining by 41% [18]. The total population of Thailand is around 68 million. Traffic law enforcement faces challenges, with minimal penalties for minor offences combined with a general cultural acceptance of breaking certain rules leading to a driving environment where potentially dangerous scenarios are commonplace. These include driving on the wrong side of the road or the pavement, jumping red traffic lights and double (or treble) parking in no-parking zones. This different driving environment could create obstacles for autonomous vehicles that are not experienced in their current testing environments.

Chiang Mai is an ancient city located in northern Thailand, the largest city outside of the central provinces. Originally founded in 1296, the current city now expands away from the moated old city with residential areas now either side of 3 ring roads encompassing much of the city. Despite new roads being constantly under construction, traffic congestion is a growing concern for residents. Within the old city, bordered by the moat, many of the streets are narrow, slow moving and one-way. There are limited public transportation options available, resulting in much of the population using private vehicles. The only mass transit option is the songthaew, a pickup truck converted to allow 2 benches of covered seating in the back. Alternatively there are a limited number of metered taxis servicing the airport, or 3-wheeled tuktuks found in the old city. None of these options are scheduled, consistent or reliable, although the regulated taxis do run on a meter. Despite strong law enforcement, the 7-day New Year holiday in 2017 resulted in the highest recorded number of road traffic accidents and related deaths in Thailand; and Chiang Mai province recorded the joint highest number of accidents and injuries. Chiang Mai therefore provides an interesting location to investigate public opinion regarding AVs [19].

2 Methodology

In 2016, a survey was conducted asking 481 residents of Chiang Mai to investigate their awareness, interest and opinions about autonomous vehicles. The survey was distributed through both online and offline means, and whilst it was a random sample, efforts were made to collect responses from different parts of the community, in order to obtain a reasonable cross-section of the society. The survey consisted of three parts: firstly some demographic data was collected, the second section collected information about existing transportation habits, including what forms of vehicle automation are already in use, and the final and largest section investigated perceptions and opinions about autonomous vehicles.

3 Results

The demographic data collected in the survey demonstrates that a broad cross section of Chiang Mai's residents were included. Female respondents made up 54% of the total, with the remaining 46% being male. The majority of respondents were under 45 years old, with 28% in the 18–25 age range, 25% in the 25–35 range, 31% in the 35–45 range, 13% in the 45–55 range and 3% in the over 55 range. Regarding the highest education levels, 11% had completed high school, while 57% had a bachelor's degree, a further 19% had a master's degree, and only 1% had a doctoral degree. Unsurprisingly 97% of those surveyed were Thai, and 3% were not. The survey was translated so respondents could answer in Thai or English.

The second part of the survey investigated the transportation habits of the sample. The commonest form of transportation was the car, which was the most popular amongst 72% of the respondents. Twenty-four percent used motorbikes the majority of the time, while just 4% used public transport. When asked which means of public transport they used on a regular basis, 47% said they do not use any public transport, while the most popular form of public transport was the songthaew, regularly used by 36%, followed by taxis (17%) and tuktuks (6%). Modern ride sharing services such as GrabTaxi and Uber were only used by 4% and 3% respectively. When asked about the current forms of vehicle automation features that they currently use, 76% use automatic transmission, 21% use automatic braking and only 7% use cruise control.

The largest part of the survey investigated the awareness and opinions concerning the use of autonomous vehicles. The first question asked whether they were aware of Google's self-driving car project, and 56% of those surveyed confirmed they had heard of it. When asked when they thought autonomous vehicles would be available, the majority expected them with 5–10 years (42%). 17% thought they would be available within 5 years, while 35% thought they would take longer to come, being available by 2035 (28%) or 2050 (7%). A further 6% felt they would never be available.

When asked when they would be ready to start using an autonomous vehicle, the responses match the technology adoption lifecycle [20]. There were a small group (8%) of early adopters (very similar to the 13% predicted by the model) who are ready to start using an AV as soon as it is available, while the early majority will be ready after a few years of testing (33%) and the late majority ready when the majority of cars are autonomous (35%). An smaller percentage of laggards (16%) said they would only be ready if they had to use an AV, while 8% didn't think they would ever be ready (Table 1).

Table 1. Opinions on AVs.

Aware of Google's self-driving car project			
Yes	56%	No	44%
How long before autonomous vehicles are available			
Within 5 years	17%	5–10 years	42%
By 2035	28%	By 2050	7%
Never	6%		
When would they be ready to start using an AV			
As soon as possible	8%	After a few years testing	33%
When the majority are AVs	35%	Only if have to	16%
Never	8%		

The next section investigated how much the respondents would be willing to pay for autonomous vehicles, the first question asking how much they would pay to convert their vehicle to an autonomous vehicle. Of those asked, 15% were willing to pay more than 100,000B for the conversion (around 3000 USD). 24% would pay in the 50,000–100,000 range, while 32% would pay in the 20,000–50,000B range and 29% would pay less than 20,000B (around 600 USD). When asked about their private vehicles, the respondents were relatively evenly split between the offered price ranges. However, when asked about renting a AV, the majority chose the cheapest price range offered with 65% saying they would pay 50B (around 1.5 USD) for a 10 KM ride. A further 25% would be willing to pay 100B (around 3 USD). A current metered taxi fare in Chiang Mai for 10 KM depending on traffic would cost around 135B.

The next section of the survey investigated the perceived benefits that might arise from AVs. Here respondents were asked to rank various criteria on a Likert scale where represented "Not Important" to 5 which was "Very Important". Most of the criteria offered were considered very important by the respondents, but the benefit of improved safety of AVs was given the highest priority (4.35). This was followed by offering access for the disabled or elderly (4.33) and then the convenience of avoiding drink driving etc. (4.26). Environmental concerns were addressed by lower vehicle emissions (4.18), improved traffic congestion (4.11) and better fuel economy (4.08). Considerations about time appeared at the bottom of the list with faster travel time and the ability to work in the car being rated on average at 3.98 (Table 2).

Table 2. Perceived benefits from AVs.

Criteria			
Better safety	4.35	Access for disabled or elderly	4.33
Convenience (drink driving etc.)	4.26	Lower vehicle emissions	4.18
Less traffic congestion	4.11	Better fuel economy	4.08
Faster travel time	3.98	Ability to work in the car	3.98

The next section investigated concerns about AVs. Again a Likert scale was offered, with 1 representing a minor concern and 5 a major concern. The results showed that the respondents had major concerns about many of the criteria offered, with "Safety" the biggest one (4.46). This was followed by security issues with hacking of the computer system ranking at 4.34, legal aspects with the liability of the driver or owner being rated at 4.29, loss of control (4.25) and interaction between AVs and conventional vehicles (4.23) followed by government regulation (4.21). The criteria that were considered less concerning were the cost of the vehicle (4.10), the speed of the vehicle (3.83) and learning to use the vehicle (3.83) (Table 3).

Table 3. Perceived concerns about AVs.

Criteria			
Safety	4.46	Hacking of computer system	4.34
Legal liability of driver/owner	4.29	Loss of control	4.25
Interaction between AVs and other vehicles	4.23	Government regulation	421
Cost of vehicle	4.10	Speed of vehicle	3.83
Learning to use vehicle	3.83		

The following section of the survey queried what the respondents would like to do while in a AV. Here the most popular activity was to look out of the window with 68% of those surveyed interested, followed by surfing the Internet (58%). Less popular activities included working (35%), sleeping (35%) and watching movies or playing games (30%) (Table 4).

Table 4. Activities in AVs.

Activity	Yes (%)	Maybe (%)	No (%)
Look out of window	68	28	5
Surf internet	58	34	9
Work	35	47	18
Sleep	35	40	25
Watch movies/Play games	30	48	22

The next section investigated where the respondents thought AVs should be used. The opinions were fairly consistent for all the locations offered, although in the countryside got the highest approval, followed by in traffic jams.

The final part of the survey asked about specific driving scenarios and the perceptions of how difficult the scenarios would be for AVs. Here again a Likert scale was used where 1 represented little perceived difficulty for AVs and 5 represented perceived difficulty. Some of these questions were directed towards local driving behaviour and etiquette. The scenario that caused the greatest concern was general bad behaviour by human drivers (3.84), followed by motorbikes riding towards the traffic (3.78). Drivers jumping red lights (3.74), was the next concern followed by unclear road signs (3.71) and then oncoming vehicles overtaking around corners (3.66). The less concerning scenarios included incorrect parking scenarios (3.24 and 3.03) and other local road users, songthaews (3.61), dogs (3.44) and pedestrians (3.35) (Table 5).

Table 5. Anticipated difficulty for AVs in specific driving scenarios.

Scenario			
General bad behaviour	3.84	Motorbikes riding towards traffic	3.78
Drivers jumping red lights	3.74	Unclear road signs	3.71
Oncoming vehicles overtaking	3.66	Songthaews	3.61
Dogs	3.44	Pedestrians	3.35
Parking in no parking zones	3.24	Double parking	3.03

4 Conclusions

The results indicate the population of Chiang Mai are fairly knowledgeable about the future of autonomous vehicles, with the majority aware of existing projects, and expecting autonomous vehicles to be available within the next 10 years, a very small minority (6%) didn't foresee a future of self-driving cars. Amongst those surveyed, the use of public transport was minimal, with nearly half never using any form of public transport, although songthaews are a public transport option. Most prefer to use their own vehicles, predominately cars, and to a lesser extent, motorbikes. This preference for personal vehicle ownership extended to the view of the future use of AVs, with a greater willingness to pay higher amounts to own their own AV than to pay per journey.

As mentioned, the willingness to adopt AVs follows a standard bell curve for new technology adoption. A group of early adopters (8%) would be keen to start using AVs as soon as they are available, while the majority of the pragmatist population would follow on after some testing and when most vehicles are autonomous. Less than 25% of those surveyed expressed reluctance to adopt the new technology. Based on this, assuming technology development continues to advance and be accredited, three quarters of the population could be willingly using their personal AVs within the next decade.

The predominant concern, and perceived benefit, was safety. The safety of AVs is considered the primary concern at this point, but it is also seen as the main benefit that might come from their use. Accessibility and convenience were the other main perceived benefits, with less interest in potential productivity, economic or timesaving benefits.

The final section of the survey concerned some of the driving conditions in Chiang Mai, and Thailand, which are partially responsible for the accident rates. There was some concern expressed about how AVs might cope with general bad driving behaviour, and incidents such as motorbikes driving towards oncoming traffic, red lights not being observed, and inappropriate overtaking. Whilst there was some concern expressed about the presented driving scenarios, the respondents were notably less concerned about these scenarios than they were about the general safety of AVs. Perhaps if AVs can allay the perceived general safety fears, then Thailand-specific driving scenarios will not present any concerns. Nonetheless it should be borne in mind that once AVs have been tested and proven in a different, cultural, driving environment, they will still need to be tested under alternative conditions.

This study demonstrates that the population of Chiang Mai are interested in the emerging AV technology, and the majority would be willing to adopt it within the next decade, largely due to perceived safety improvements. Most would prefer to own their own AVs rather than sharing them, as public transport remains unpopular.

References

1. Anderson, J.M., Nidhi, K., Stanley, K.D., Sorensen, P., Samaras, C., Oluwatola, O.A.: Autonomous Vehicle Technology: A Guide for Policymakers. RAND Corporation, Santa Monica (2014)
2. Fagnant, D.J., Kockelman, K.: Preparing a nation for autonomous vehicles: opportunities, barriers and policy recommendations. Transp. Res. Part A **77**, 167–181 (2015)
3. Zakharenko, R.: Self driving cars will change cities. Reg. Sci. Urban Econ. **61**, 26–37 (2016)
4. Zhang, W., Gubathakurta, S., Fang, J., Zhang, G.: Exploring the impact of shares autonomous vehicles on urban parking demand: an agent-based simulation approach. Sustain. Cities Soc. **19**, 34–45 (2015)
5. Romem, I.: How will driverless cars affect our cities? (2013). http://www.cityminded.org
6. Rajasekhar, M.V., Jaswal, A.K.: Autonomous vehicles: the future of automobiles. In: Transportation electrification conference, ITEC, IEEE International (2015)
7. Firnkorn, J., Müller, M.: Selling mobility instead of cars: new business strategies of automakers and the impact on private vehicle holding. Bus. Strateg. Environ. **21**(4), 264–280 (2012)
8. Krueger, R., Rashidi, T.H., Rose, J.M.: Preferences for shared autonomous vehicles. Transp. Res. Part C **69**, 343–355 (2016)
9. Bansal, P., Kockelman, K.M.: Forecasting Americans' long-term adoption of connected and autonomous vehicle technologies. Transp. Res. Part A **95**, 49–63 (2017)

10. Bansal, P., Kockelman, K.M., Singh, A.: Assessing public opinions of and interest in new vehicle technologies: an Austin perspective. Transp. Res. Part C **67**, 1–14 (2016)
11. Kyriakidis, M., Happee, R., de Winter, J.C.F.: Public opinion on automated driving: results of an international questionnaire among 5000 respondents. Transp. Res. Part F Traffic Psychol. Behav. **32**, 127–140 (2015)
12. Choi, J.K., Ji, Y.G.: Investigating the importance of trust on adopting an autonomous vehicle. Int. J. Hum. Comput. Interact. **31**(10), 692–702 (2015)
13. Chadbunchachai, W., Suphanchaimaj, W., Settasatien, A., Jinwong, T.: Road traffic injuries in Thailand: current situation. J. Med. Assoc. Thai. **95**(Suppl. 7), S274–S281 (2012)
14. World Health Organisation: WHO report 2015: global status report on road safety (2015)
15. Porapakkham, Y., Rao, C., Pattaraarchachai, J., Polprasert, W., Vos, T., Adait, T., et al.: Estimated causes of death in Thailand, 2005: implications for health policy. Popul. Health Metr. **8**, 14 (2010)
16. Tongklao, A., Jaruratanasirikul, S., Sriplung, H.: Risky behaviour and helmet use among young adolescent motorcyclists in southern Thailand. Traffic Inj. Prev. **17**(1), 80–85 (2016)
17. Berecki-Gisolf, J., Yiengprugsawan, V., Kelly, M., McClure, R., Seubsman, S., Sleigh, A.: The impact of the Thai motorcycle transition on road traffic injury: Thai cohort study results. PLoS ONE **10**(3), e0120617 (2015)
18. ASEAN-JAPAN Transport Partnership: Data template for road transport (2013)
19. Bangkok Post: 2017 death toll highest on record. http://www.bangkokpost.com/news/general/1174796/2017-death-toll-highest-on-record. Accessed 5 Jan 2017
20. Moore, G.: Crossing the chasm, marketing and selling disruptive products to mainstream customers. Collins Business Essentials (2014)

A Study of Conceptual Connected Communication Vehicular Network Using Distributed Cluster Algorithm

Han-Chun Song and Jinhong Kim(✉)

Department of Information and Communication Engineering,
Seoil University, Seoul, Korea
{sanho, jinhkm}@seoil.ac.kr

Abstract. This research paper proposes a Connected Vehicular Communication Network using Distributed Cluster Algorithm (DCA). We are introduced to novel cluster metric, network model. Especially, we propose vehicular communication network on cluster algorithm against changes in environmental changed parameters, and P2P (Point to Point) of a graph such as ring search, neural network and so on. In addition, we use the localized notion of node in conjunction with a universal link metric to derive a DCA for the vehicular network. Moreover, we design a distributed vehicular network that the proposed DCA forms a more robust cluster structure.

Keywords: Distributed Cluster Algorithm · Vehicular Communication Network

1 Introduction

VANET (Vehicular Ad hoc Networks) are employed by intelligent transport systems to operate wireless communications in the vehicular environments [1–3]. VANET are designed to provide a reliable and safe environment for drivers. Their derivers could be informed of situations by vehicular communications and exchanging the own information about surrounding environments. The vehicles in VANET are similar to the mobile nodes in the MANET [4–6]. However, VANET inherit most of the characteristics of MANET, but VANET have some unique characteristics such as high rate of topology changes, and high density of the network, and so on [7, 8]. For our research have the new strategies and considering, we have to propose the congestion in Vehicular Communication network with DCA.

2 Vehicular Communication and System

The main of V2V (Vehicle to Vehicle) communication is to standardize the protocols and interfaces of wireless communication between the vehicles and their environment [9]. It makes different vehicle as well as communicate with access points or the road

K.J. Kim et al. (eds.), *IT Convergence and Security 2017*,
Lecture Notes in Electrical Engineering 450,
DOI 10.1007/978-981-10-6454-8_2

side units. V2V communications form a decentralized network that is well suited for active safety applications. Apart from safety applications, gathered information helps in traffic management to support traffic flow [10–12]. Especially, V2V communication systems are including In-Vehicle Domain, Ad-Hoc Domain, and Infrastructure Domain. In order to enable vehicles and the corresponding infrastructure to exchange data in adequate manner by Radio System [12–14]. There are two types of communication channels used by V2V radio system. (1) ***Dedicated V2V Channels*** have network control and critical safety applications, and road safety and traffic efficiently applications. (2) This system provide ***public channels as specified in IEEE 802.11 a/b/g***. In Fig. 1 shows a cross-layer control architecture in VANET, this is considered for detecting and controlling between some information from the application layer and detection by sensing the channel in the physical layer. In addition, the network layer can control by routing that efficiently rebroadcast the message. The prioritizing and scheduling message at MAC layer can significantly help control in VANET [15].

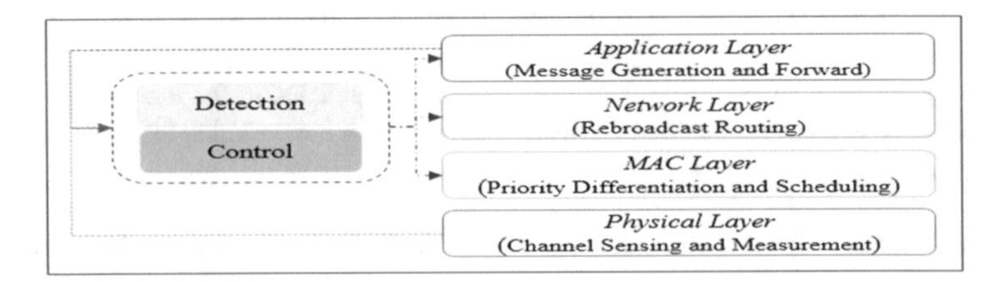

Fig. 1. Cross-layer control architecture in VANET

3 Proposed System

3.1 C-VCN (Connected Vehicular Communication Network) Model

We design that WLAN (Wireless LAN) is based on IEEE 802.11 standard and WWAN (Wireless WAN) is implemented by 3G/4G by hybrid network model. Wireless network is constituted by different layers, and each layer has its own responsibilities. WWAN has already had some acknowledge standards. The most popular one is 3G/4G network. In our research model, 3G/4G wireless network as WWAN part. Each vehicle communicates with its BaseStation (BS). As shown in Fig. 2, BaseStation access to internet through wired network. Generally, 3G/4G network provides relative low throughput with large cell coverage. Our C-VCN model represents vehicular communication network architecture. The involved entities of the network architecture are following Fig. 2 as bellows:

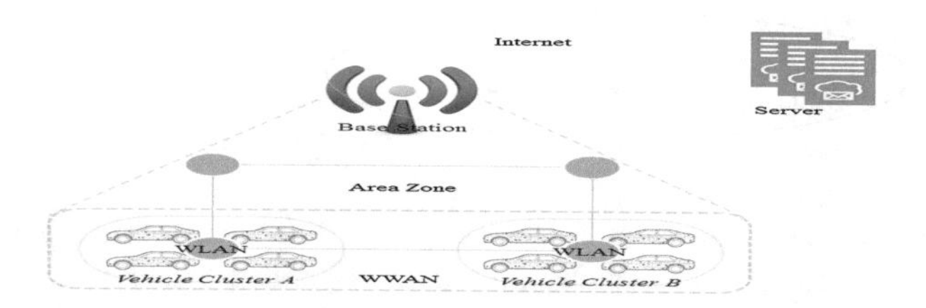

Fig. 2. Vehicular network architecture for connected vehicular communication network. Area zone nodes are vehicle running and in operating network. Communication in vehicle provides an IEEE 802.21 interface and a 3G/4G interface. Base Station is central infrastructure in 3G/4G Cellular Network, and Servers are functional entities that provide service to these clients.

3.2 Connected Communication Vehicular Network (CCVN)

In vehicular communication network, cluster structure is implemented by the cluster algorithm. This support an appropriate topology for distribute node and decide the effect of running algorithm. Our concept is a highly dynamic wireless network, and is motivated by the definition of network capacity. We need to be solved clustering metrics by clustering. This metrics would provide the quantitative definition, and should reflect the characteristic of their network. In general, network topology in VANET is frequently changing. In our simple scenario, considering vehicular network contain V_n nodes. If it takes an arbitrary time, the network topology could be describes as a graph $G = \{V, E\}$, where $V_n = (V_1, V_2, V_3, V_{4,....} V_n)$ is the set of nodes and E is the set of links between nodes. In a graph $G = \{V, E\}$, the weight matrix is $W = \{w_{ij}\}^n$. Let the matrix M have $m_{ij} = \sum_{k=1}^{n} w_{ik}$ $(i = j)$, *otherwise 0.* M is the degree matrix for weight W. According to this graph, a cluster is named community, and so vehicular clustering can be classified into a community detection problem, which focuses on aggregating similar nodes together with the given features of nodes.

When we assume the following properties for this graph, Fig. 3 is shown the abstraction vehicular network model. We centrality measure for both nodes and links are quantifies the robustness of a network graph with the environmental changes, traffic events, topology modifications, and destination for traffic. After all, VANET is a connected graph, a random node occurs when travels between nodes.

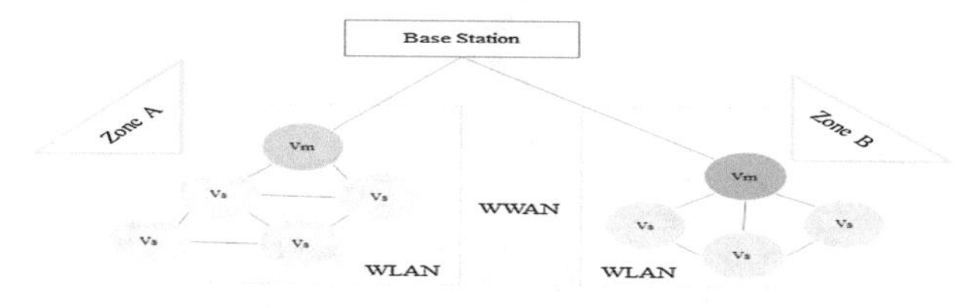

Fig. 3. Abstraction vehicular network model. G is an undirected graph. For each link by zone, this has a weight w which is defined to represent between node V_m (*Vehicle Master*) and node V_s (*Vehicle slave*). Each node is communicated with V2V, so position and velocity of each node. In addition, each node has the same transmission in WLAN area, is composed equally in the proposed cluster algorithm.

3.3 Distributed Cluster Algorithm for CCVN

In our research paper, the proposed distributed cluster algorithm is designed for the purpose of each vehicles trajectory in CCVN. This algorithm assumes each vehicle is aware of their location and velocity using Base Station. Each vehicle inside vehicle cluster provides their context from the target by Fig. 2. In other words, their location is reached on vehicle distance that can be used in this algorithm to acquisition information such as enable trajectory. ***Algorithm Description***. This is divided into three cases 'initialization', 'cluster maintenance', and 'trajectory'. First, in case of initialization cluster is created and the initial cluster (V_m) is selected by '*head*'. Second, in case of processing, in each vehicle (V_s) node performs their different tasks for cluster maintenance, and recommend in changed trajectory in each cluster. Base Station (BS) broadcast a control message to the entire network the vehicle information such as context, condition, and situation and waits to receive response message from each vehicle or cluster. For BS waits a response message, it has a plan to decide trajectory policy by considerable environment from each vehicle or cluster. When BS receives a response message from them, send to specific zone in the network if it has a rough about the context information. BS performs procedure from Algorithm 1 and show Fig. 4

Algorithm 1 BaseStation Procedure

Describe the Function.
SendBS (Node ID, PacketId, TrajectoryInfo, CurTime);

Task Action.
IF do not received V_m (Head) from the V_s ( ) then,
  sendBS(NodeID, PacketId, TrajectoryInfo, CurTime)
end if
IF received V_m (Head) from the V_s ( ) then,
  sendBS(new NodeID, new PacketId, new TrajectoryInfo, new CurTime)
end if

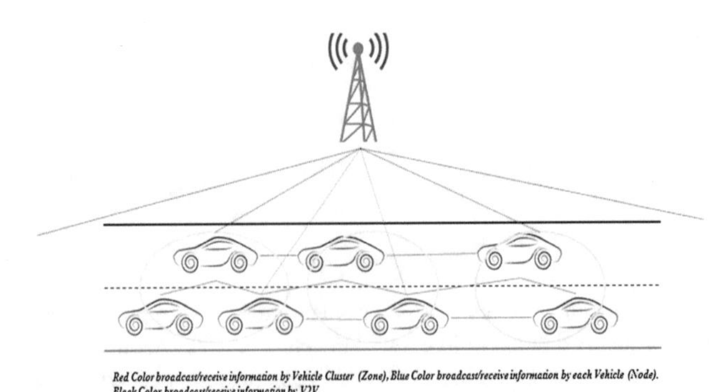

Fig. 4. BaseStation Communication Function. It is a central/individual broadcasting and receiving with trajectory for cluster and particular vehicle.

In DCA, V_m is the central management V_s in which all V_s had a role in managing in the cluster. Then, our algorithm is considered a central algorithm. The V_m is responsible for cluster maintenance in order to make a reliable trajectory. Accordingly, we have solved by reliable trajectory with one/more candidate V_m of another clusters. The cluster maintenance function of V_m is shown in Algorithm 2. The V_m begin to send an information message for their tasks by communication inside cluster, and V_m supposed to be sent at each ΔT_R time interval. Also, we show a Fig. 5.

Algorithm 2 V_m in cluster Procedure

Task Action

While (Trajectory == true) do

send V_m (Node ID, PacketId, TrajectoryInfo, CurTime);

$V_m \leftarrow V_s$

Calculate ΔT_R ( )

sendV_m ( );

………..

If receive V_m ( ) then,

Update.information (NodeID, PacketId, TrajectoryInfo, CurTime)

endIf

If do not received $V_m \leftarrow V_s$ ( ) then,

Search.Trajectory.info (new NodeID, new PacketId, new TrajectoryInfo,new CurTime)

endIf …..

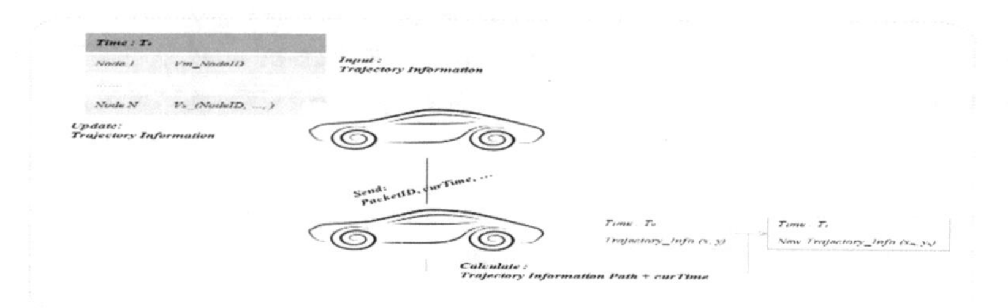

BSM (Base Station Message)

NodeID	PacketID	TrajectoryInfo	curTime

Vm (Master Vehicle in Cluster Message)

NodeID	PacketID	curTime	Other Information

Vs (Slave Vehicle in Cluster Message)

NodeID	PacketID	curTime	Node_TrajectoryInfo	Other Information

Fig. 5. V_m and V_s broadcast/communicate transaction on WLAN and message information

4 Experiment and Evaluation

We have experiment and test the proposed algorithm and from Figs. 2, 3, 4 and 5 with our scenario to show the effect of trajectory on cluster. As well, we have performance of each algorithm which is communicated for trajectory purpose between V_m and V_s. Although we had a weak point of simulation area, this research used networking technology for vehicle trajectory and communication purpose which may be a future advancement by many researchers. This experiment result of our algorithm is shown in Table 1.

Table 1. Experiment Considerable Elements

Parameter	Value
Experiment environment	Gyeongbu express highway, Korea
Number of vehicle (node)	50, 100, 150, 200
Data packet size	1000 byte
Data packet frequency	0.5 Hz
Control packet frequency	1 Hz
Transmission rate	1 Mbps
Communication range	50, 100, 250, 500 m
Traffic type	UDP
Mac protocol	IEEE 802.11

We have experimented our algorithm with vehicle numbers to the effect of network density on clustering performance, and we had a velocity range (25–35 m/s) and transmission range (100 m). We shows the effect number of vehicles on the V_m metric. In our algorithm, we have considered a threshold for changing the V_m. This threshold is defined in a way to decrease changes in some cases. The evaluation result in Fig. 6 shows that increasing the number of vehicle have a positive effect on the V_m. The good

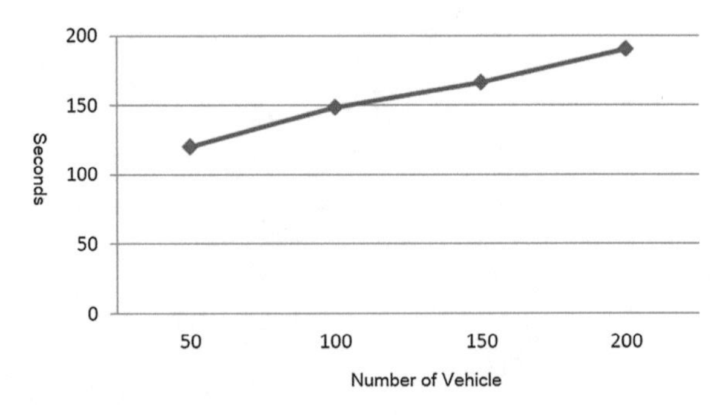

Fig. 6. Increasing the number of vehicle on the V_m.

reason is appropriate V_m metric which is not affected so much by cluster structure changes, and is a vehicle with the most similar movement pattern to the trajectory.

Figure 7 shows the effect of network density changes on packet delivery ratio. In dense networks more vehicles are capable of detecting the trajectory. Accordingly, the number of cluster increase which results in more data message transmission in the cluster. After all, as the number of message increase, the probability of packet collision increases, and packet delivery ratio falls down as a result.

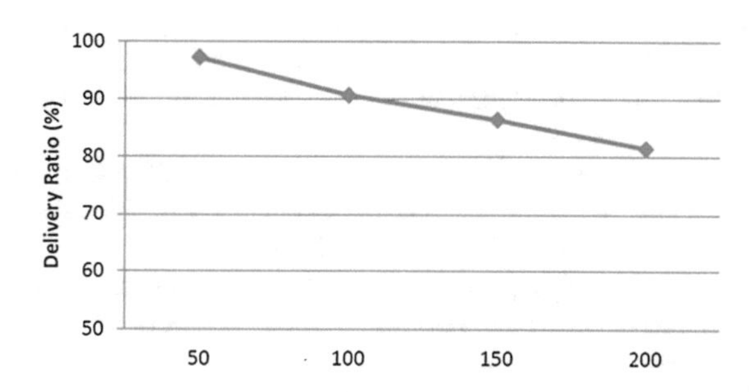

Fig. 7. Packet Delivery Ratio with the effect of network density

5 Conclusions

The vision of V2V Communication in general interested most of the researchers in the automotive industry. Their standardization activities are based on the IEEE 802.11 topology, and will have an opportunity to exchange information among themselves, road sensors and traffic signs by vehicular communication network. In this research paper a small prototype communication network model in highway environment is presented, and also proposed a distributed cluster algorithm for area of vehicular network. Considering the highly dynamic nature of the vehicular network, we emphasize the robustness of the resulting clustering. Network criticality is a global metric that quantifies the robustness of network topology against the variable environment in the network topology. In addition, in the proposed clustering by Fig. 2 that we derived new concept with DCA. For the distributed case, we localize the connection of V2V network and use as a metric for DCA. DCA significantly improves the time of clusters and state change times for vehicle trajectory. It is more suitable for mobile wireless networks such as multi-hop, Ad-hoc network and so on, but our research prototype is limited vehicular communication environment in real world. Nevertheless, we are anticipated in developing in future applications and expansion for vehicle infrastructure.

Acknowledgement. The present research has been conducted by the Research Grant of Seoil University in 2017.

References

1. Thakkar, A., Kotecha, K.: Cluster head election for energy and delay constraint applications of wireless sensor network. IEEE Sens. J. **14**(8), 2658–2664 (2014)
2. Zhang, Z., Boukerche, A., Pazzi, R.: A novel multi-hop clustering scheme for vehicular ad-hoc networks. In: Proceedings of the 9th ACM international symposium on mobility management and wireless access, pp. 19–26. ACM, Miami (2011)
3. Tizghadam, A., Leon-Garcia, A.: Autonomic traffic engineering for network robustness. IEEE J. Sel. Areas Commun. **28**(1), 39–50 (2010)
4. Bo, L., et al.: Component-based license plate detection using conditional random field model. IEEE Trans. Intell. Transp. Syst. **14**(4), 1690–1699 (2013)
5. Barani, H., Fathy, M.: An algorithm for localization in vehicular ad-hoc networks. J. Comput. Sci. **6**(2), 168 (2010)
6. Hajiaghajani, F., et al.: HCMTT: hybrid clustering for multi-target tracking in wireless sensor networks. In: IEEE International Conference on Pervasive Computing and Communications Workshops (PERCOM Workshops) (2012)
7. Chunhua, Z., Cheng, T.: A K-hop passive cluster based routing protocol for MANET. In: 5th International Conference on Wireless Communications, Networking and Mobile Computing, WiCom 2009 (2009)
8. Visintainer, F., D'Orazio, L., Darin, M., Altomare, L.: Cooperative systems in motorway environment: the example of trento test site in Italy. In: Advanced Microsystems for Automotive Applications. Lecture Notes in Mobility, pp. 147–158. Springer, Berlin (2013)
9. Hua, Q., Zhang, W.: Charging scheduling with minimal waiting in a network of electric vehicles and charging stations. In: ACM VANET, Las Vegas (2011)
10. Asgari, M., Jumari, K., Ismail, M.: Analysis of routing protocols in vehicular ad hoc network applications. In: Software Engineering and Computer Systems, vol. 181, pp. 384–397. Springer, Heidelberg (2011)
11. Prakash, A., Tripathi, S., Verma, R., Tyagi, N., Tripathi, R., Naik, K.: A cross layer seamless handover scheme in ieee 802.11p based vehicular networks. In: Contemporary Computing, vol. 95, pp. 84–95. Springer, Heidelberg (2010)
12. Chiu, K.L., Hwang, R.H., Chen, Y.S.: Cross-layer design vehicle-aided handover scheme in VANETs. Wireless Communications and Mobile Computing **11**, 916–928 (2009)
13. Kim, J., Kim, S.: A study of multi-hop wireless networks model for smart vehicle adaptive traffic. Int. J. Multimed. Ubiquitous Eng. **8**(6), 361–366 (2013)
14. Kim, J.-S., Kim, J.-H.: Vehicle trajectory discovery for vehicular wireless networks. Int. J. Smart Home **8**(1), 157–164 (2014)
15. Kim, J.-H., Lee, E.-S.: KSVTs: towards knowledge-based self-adaptive vehicle trajectory service. In: Information Technology Convergence, vol. 253, No. 1, pp. 387–393. Springer, Dordrecht (2013)

Healthcare and Wellness

EEG Based Classification of Human Emotions Using Discrete Wavelet Transform

Muhammad Zubair[1(✉)] and Changwoo Yoon[2]

[1] Korea University of Science and Technology, Daejeon, Korea
zubair5608@gmail.com
[2] Electronics and Telecommunication Research Institute, Daejeon, Korea

Abstract. Electroencephalography is widely used to study the dynamics of neural information processing in the brain and to diagnose brain disorder and cognitive processes. In this paper, we proposed EEG based emotion recognition system using Discrete Wavelet Transformation. A set of highly significant features based on wavelets coefficients has been extracted which also includes modified wavelet energy features. In order to minimize redundancy and maximize relevancy among features, mRMR algorithm is significantly applied for feature selection. Multi class Support Vector Machine is used to perform classification of four classes of human emotions. EEG recordings of "DEAP" database are used in this experiment. The proposed approach shows significant performance compared to existing algorithms.

Keywords: EEG signals · Emotion recognition · Wavelet transform · SVM

1 Introduction

Automatic detection and recognition of different emotional states is a salient topic in the fast growing research field of affective computing. Emotions are complex states of minds comprised of numerous psychophysiological components, such as bodily changes, cognitive reactions and thoughts. Numerous computational models and algorithms for automatic recognition of emotions have been provided by Affective computing, which integrates knowledge of computer science, physiology, artificial intelligence and biomedical engineering.

Emotions affect all aspects of our daily life and has significant influence on our health. State of depression, anxiety and anger disrupt human immune system and thus associated with many chronic diseases. Bringing into play the current advances in IOT and sensor networks [1], smart healthcare systems should be introduce to improve overall quality of life. The development of automatic emotion recognition system can be very useful in regulating self-emotions and would revolutionize applications in education, entertainment and security.

Philosophers and psychologists presented various theories of emotions. Based on these theories, numerous methods have been developed to detect and recognize different emotions using facial images [3], speech signals [4], gestures and physiological signals.

K.J. Kim et al. (eds.), *IT Convergence and Security 2017*,
Lecture Notes in Electrical Engineering 450,
DOI 10.1007/978-981-10-6454-8_3

Physiological signals including EEG signals are considered to be the most useful signals for human emotion recognition due to its strong correlation with emotions and independence of people's will. Facial images based emotion recognition have a major flaw of suppressing and intentional control of emotions while physiological signal originate from Autonomous Nervous System activity, cannot be controlled intentionally. Experimental evidence shows that physiological/bio signals can be influence by the activity of Autonomic Nervous System (ANS) and can convey information regarding human emotion [5, 6]. In this paper, we proposed human emotion recognition system using wavelet energy feature along with statistical and modified wavelet energy feature in order to improve the performance of emotion recognitions systems.

This paper is organized as follows: Sect. 2 provides a review on different factor of emotion recognition system. Section 3, illustrates the detailed methodology including dataset description, procedure for feature extraction, feature selection and classification. Performance of the proposed method and concluding remarks are given in Sects. 4 and 5 respectively.

2 Related Work

EEG based emotion recognition has gain a lot of interest and different emotion classification system have been proposed by the researchers. The results of these systems highly depends on five basic factors which includes number of participants, stimulus, emotion modeling, feature extraction and classifier. Different techniques have been employed by the researchers to investigate these factors and thus these emotion classification systems cannot be compared. However a short review of the recent work done is presented in this section.

Emotion modeling: Emotion is a mental state or feeling that arises involuntarily and comprised of different components such as feelings, bodily changes, behavior and thoughts. In literature numerous emotion models have been proposed. However the two most utilized categories are Discrete Emotional Models (DEM) and Affective dimensional model (ADM). DEM deals with six basic universal categories of emotions: happiness, surprise, anger, disgust, sadness and fear [2]. ADM deals with the description

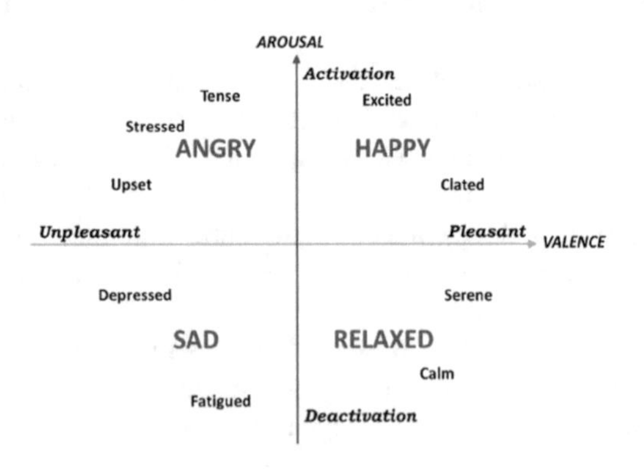

Fig. 1. Circumplex model of emotions.

of emotions in some coordinate system. It characterize emotion into two affective parameters, Arousal and Valence. The most commonly used dimensional model is Circumplex Model of Affects (CMA) [7] as given in Fig. 1.

Emotion elicitation: There are numerous methods of emotion elicitation. The most widely used methods for emotion induction includes images, video clips and sound clips etc. The most popular existing databases are: IAPS—the International Affective Picture System and IADS—the International Affective Digitized Sound System facilitate the task of emotion recognition. IAPS [8] and IADS [9] provides a collection of stimuli publically to researchers in the study of emotion.

Feature extraction and classification: Different characteristics of EEG signals can be captured and used as features for classification of emotions. These features can be placed in one of two domain, time domain and frequency domain. Time domain feature can be extracted from raw EEG signal and includes mean, standard deviation etc. [10]. Frequency domain features include the power of different frequency bands of EEG signals. In addition to these, several other features extraction techniques have been presented in the literature which includes High Order Crossings [10], Discrete Wavelet Transformation [11], fractal dimensions [12] and Independent Component Analysis [13]. For classification task, several machine learning algorithms have been used as classifiers. These classifiers includes Support Vector machine [14], Neural Networks [15] and Quadratic Discriminant Analysis [16].

3 Methodology

3.1 Dataset

Recent advances in emotion recognition have increased the interest of many researchers and encouraged them to create databases containing visual, speech and physiological emotion data. These databases includes MAHNOB-HCI [17] and DECAF [18] etc. In this study, we used a public available database called DEAP proposed by Koelstra et al. [19]. 32 healthy participants (16 male and 16 female), aged between 19 and 32, took part in the experiment. The EEG and peripheral physiological signals which includes electrocardiogram, galvanic skin response, respiration, skin temperature, blood volume, electromyograms (EMG) and electrooculogram (EOG) were recorded from these subjects while watching 40 different music videos. Biosemi Active Two system was used to record EEG signals over the scalp from 32 electrodes according to the international 10–20 system as shown in Fig. 2. Preprocessing of EEG signals was performed in order to denoise the heavily distorted EEG signals from motion artifacts, EOG artifacts due to eye blinking and power supply noise. Initially EEG signals were recorded with 512 Hz sampling frequency which were down sampled to 128 Hz during preprocessing. A bandpass frequency filter from 4.0–45.0 Hz was applied to filter EEG signals. The elimination of EOG artifacts was performed using blind source separation technique. In the last of experiment every subject performed self-assessment to evaluate and rate the emotional state caused by each video using Self-Assessment Manikins. Rating was performed by participant for each music video in term of levels of arousal, valence, like/dislike, dominance and familiarity. In this paper, we used the preprocessed data released by Koelstra et al. [19].

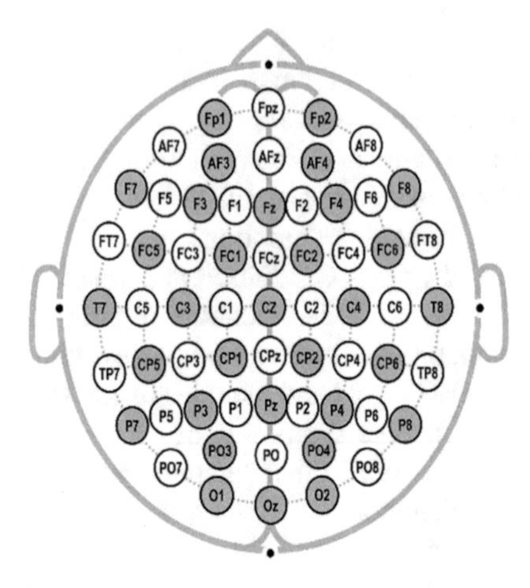

Fig. 2. International 10–20 system for 32 electrodes (Marked as gray)

3.2 Feature Extraction

Statistical-Based Features

Due to the nonlinear nature of EEG signals and brain complexity, nonlinear feature like high order crossing and fractal dimension are widely employed by researchers in recent publication. However, simple features like mean, standard deviation and band power are still considered beneficial for emotion recognition system. For 32 channel EEG data provided in DEAP dataset, we extracted statistical features in combination with wavelet based feature to improve emotion recognition accuracy. These features include

- Mean of the raw signal

$$\mu_X = \frac{1}{N}\sum_{n=1}^{N} X(n) \tag{1}$$

- The standard deviation of the raw signal

$$\sigma_X = \sqrt{\frac{1}{\mathrm{N}}\sum\nolimits_{\mathrm{n}=1}^{\mathrm{N}} \left(\mathrm{X(n)}-\mu_{\mathrm{X}}\right)^2} \tag{2}$$

- The mean of the absolute values of the first difference of the raw signal

$$\delta_X = \frac{1}{N-1}\sum_{n=1}^{N-1} |X(n+1) - X(n)| \tag{3}$$

- The mean of the absolute values of the first signal of the standardized signal

$$\overline{\delta_X} = \frac{1}{N-1}\sum_{n=1}^{N-1}|\overline{X}(n+1) - \overline{X}(n)| = \frac{\delta_X}{\sigma_X} \tag{4}$$

- The mean of the absolute values of the second difference of the raw signal

$$\gamma_X = \frac{1}{N-2}\sum_{n=1}^{N-2}|X(n+2) - X(n)| \tag{5}$$

- The mean of the absolute values of the second difference of the standardized signal

$$\overline{\gamma_X} = \frac{1}{N-1}\sum_{n=2}^{N-2}|\overline{X}(n+2) - \overline{X}(n)| = \frac{\gamma_X}{\sigma_X} \tag{6}$$

Wavelet-Based Features

Discrete Wavelet Transform is a powerful analytical tool for non-stationary signals and is widely used for time-frequency analysis of EEG signals due to its non-stationary nature. DWT decomposes EEG signal into different frequency bands with successive high pass and low pass filters. The high pass filter gives detail coefficients while low pass filter gives approximation coefficients. In this paper, we used Daubechies Wavelet Transform (db4) coefficients which is considered best for multiresolution analysis of EEG signals and EEG signal with 128 Hz sampling rate. Five levels Daubechies wavelet of order 4 is applied for decomposition of EEG signals into five frequency bands, delta, theta, alpha, beta and gamma as given in the Table 1.

Table 1. EEG signals decomposition into different frequency bands

Frequency	Decomposition level	Frequency band	Frequency bandwidth (Hz)
0–4	A5	Delta	4
4–8	D5	Theta	4
8–16	D4	Alpha	8
16–32	D3	Beta	16
32–64	D2	Gamma	32

After Discrete Wavelet Transformation, we estimated wavelet energy and wavelet entropy according to Eqs. 1 and 2 respectively.

$$E_l = \sum_{n=1}^{2^{S-1}-1}|C_X(l,n)|^2, N = S^2, 1 < l < S \tag{7}$$

$$E_l = \sum_{n=1}^{2^{S-1}-1}|C_X(l,n)|^2 \log(|C_X(l,n)|^2), N = S^2, 1 < l < S \tag{8}$$

Where $C_X(l, n)$ Wavelet coefficients associated with all five sub-bands are used to estimate wavelet energy and wavelet entropy. Another energy based feature set proposed by Murugappan et al. [11] is used in this paper. These features includes Recoursing Energy efficiency (REE), Logarithmic Recoursing Energy Efficiency (LREE) and Absolute Logarithmic Recoursing Energy Efficiency (ALREE). These features are estimated for gamma band as follows.

$$REE_{gamma} = \frac{E_{gamma}}{E_{total}} \tag{9}$$

$$LREE_{gamma} = \log_{10}\left[\frac{E_{gamma}}{E_{total}}\right] \tag{10}$$

$$ALREE_{gamma} = \left|\log_{10}\left[\frac{E_{gamma}}{E_{total}}\right]\right| \tag{11}$$

3.3 Feature Selection and Classification

Feature selection is performed in order to mitigate the high dimensionality feature space problem. In this step, the most suitable subset of all derived features is selected which not only solve the problem of dimensionality but also increase the classification accuracy due to the reduction of noise caused by irrelevant features. In this paper, successfully applied maximum relevancy and minimum redundancy algorithm (mRMR) for feature selection. After selection of most relevant features, classification is performed using machine learning classifier. For this purpose a multi class Support Vector Machine is used with radial basis function.

4 Experimental Results

The EEG recordings of 32 subjects of DEAP database have been used to classify four main classes. The arousal and valence scores on the scale from 0 to 9 is mapped into two levels, high and low. The resulting four classes are, high arousal/high valence (HAHV), high arousal/low valence (HALV), low arousal/high valence (LAHV) and low arousal/low valence (LALV). For performance evaluation of the system, EEG data is divided into two portions, training data and test data. 70% of the total data is used for training purpose while the remaining 30% was used for testing. For classification, two machine learning algorithms, Support Vector Machine and Quadratic discriminant analysis are used to classify EEG data into four classes. Grid search approach was adopted for parameter optimization. In this experiment, best performance is given by SVM with overall accuracy of 49.7% using all channels data.

In order to implement a less complex and user friendly emotion recognition method, we reduced the numbers of EEG channels as much as possible. For this purpose, we selected a group of 15 EEG channels namely Fp1, Fp2, AF3, F3, F4, F7, F8, P7, O1, O2, P8, CP3, CP4, C4 and C3, that belongs to all four major lobs of the brain. Research shows that, left frontal lobe and right frontal lobe exhibit certain activity when a negative or positive emotion is experienced by a person [10]. Furthermore, the related research also reveals that different sub bands (Delta, theta, Alpha, Beta and Gamma) are activated by different emotional states in specific brain regions [20]. In this paper, we also investigated the activity of all five bands of EEG signals in selected channels for the aforementioned four classes of emotion. The classification of emotions is performed using each frequency band separately for combination of different sets of channels. EEG signal acquired from frontal lobe (FP1, FP2, F3 and F4) and temporal lobe (T7, T8) showed best performance with features extracted from Gamma band and achieved an overall classification accuracy of 48.8% for four classes of emotions which is close to the prior accuracy gained using all channels data of EGG signals (Table 2).

Table 2. Classification accuracy

Method	HAHV	HALV	LAHV	LALV	Overall
QDA	44.4%	47.7%	45.1%	44.6%	45.4%
SVM	52.1%	49.1%	49.6%	48.3%	49.7%

5 Conclusion

In this paper, we presented emotion recognition system using the most significant features set extracted from coefficients of Discrete Wavelet Transformation. A public available database called DEAP has been in this work. The EEG recordings of 32 participants have been utilized to extract statistical based feature and wavelet based feature. A feature selection algorithm was adopted to select the most significant and relevant features in order to mitigate the problem of dimensionality, irrelevancy and redundancy.

The proposed approach can significantly classify four classes of emotions using Support Vector Machine from which following can be concluded. First, feature extracted using Discrete Wavelet Transformation effectively represent emotional state of the users. Second, Gamma band holds rich information of all four classes of emotions. Third, we found that there is a strong correlation in frontal brain region related to Gamma band for all four classes of emotions which validates the role of frontal lobe in emotion recognition. In future, research will be conducted on fusion of EEG signals with others physiological signals for high performance.

Acknowledgement. This work was supported by Institute for Information & communications Technology Promotion (IITP) grant funded by the Korea government(MSIP) (Development of SW fused Wearable Device Module and Flexible SW Application Platform for the integrated Management of Human Activity).

References

1. Bilal, M., Kang, S.-G.: An authentication protocol for future sensor networks. Sensors **17**(5), 979 (2017)

2. Ekman, P.: Emotions Revealed. Times Books (2003)
3. Zhang, L., Tjondronegoro, D.: Facial expression recognition using facial movement features. IEEE Trans. Aff. Comput. **2**, 219–229 (2011)
4. Wang, K., Ning, A., Li, B.N., Zhang, Y.: Speech emotion recognition using Fourier parameters. IEEE Trans. Aff. Comput. **6**, 69–75 (2015)
5. Anttonen. J., Surakka, V.: Emotions and heart rate while sitting on a chair. In: Proceedings of the SIGCHI Conference on Human Factors in Computing Systems, pp. 491–499 (2005)
6. Jones, C.M., Troen, T.: Biometric valence and arousal recognition. In: Proceedings of the 19th Australasian Conference on Computer-Human Interaction, pp. 191–194 (2007)
7. Posner et al.: The circumplex model of affect: an integrative approach to affective neuroscience, cognitive development, and psychopathology. Dev. Psychopathol. **17** (2005)
8. Lang, P.J., Bradley, M. M., Cuthbert, B.N.: International affective picture system (IAPS): affective ratings of pictures and instruction manual. Technical repory A-8 (2008)
9. Bradley, M.M., Lang, P.J.: The international affective digitized sounds (iads-2): Affective ratings of sounds and instruction manual. University of Florida, Gainesville, FL, USA, Technical report B-3 (2007)
10. Petrantonakis et al.: Emotion recognition from EEG using higher order crossings. IEEE Trans. Inf. Technol. Biomed. **14**(2) 186–197 (2010)
11. Murugappan et al.: Classification of human emotion from EEG using discrete wavelet transform. J. Biomed. Sci. Eng. **3**(4), 390 (2010)
12. Sourina, O., Yisi, L.: A fractal-based algorithm of emotion recognition from EEG using arousal-valence model. BIOSIGNALS (2011)
13. Lan, T., et al.: Estimating cognitive state using EEG signals. In: 2005 13th European IEEE Signal Processing Conference (2005)
14. Koelstra, S., et al.: Single trial classification of EEG and peripheral physiological signals for recognition of emotions induced by music videos. In: Proceeding of the International Conference on Brain Informatics, BI 2010, Toronto, Canada, pp. 89–100 (2010)
15. Wijeratne, U., et al.: Intelligent emotion recognition system using electroencephalography and active shape models. In: Proceedings of the IEEE EMBS Conference on Biomedical Engineering and Sciences, IECBES 2012, pp. 636–641 (2012)
16. Khalili, Z., Moradi, M. H.: Emotion recognition system using brain and peripheral signals: using correlation dimension to improve the results of EEG. In: Proceedings of the International Joint Conference on Neural Networks, IJCNN 2009, Atlanta, pp. 1571–1575 (2009)
17. Soleymani, M., et al.: A multimodal database for affect recognition and implicit tagging. IEEE Trans. Aff. Comput. **3**(1), 42–55 (2012)
18. Abadi, M., et al.: DECAF: MEG-based multimodal database for decoding affective physiological responses. IEEE Trans. Aff. Comput. **6**(3), 209–222 (2015)
19. Koelstra, S., et al.: Deap: a database for emotion analysis; using physiological signals. IEEE Trans. Aff. Comput. **3**(1), 18–31 (2012)
20. Daimi, S.N., Saha, G.: Classification of emotions induced by music videos and correlation with participants rating. Expert Syst. Appl. **41**(13), 6057–6065 (2014)

Real-Time Posture Correction Monitoring System for Unconstrained Distraction Measurement

Ji-Yun Seo[1], Yun-Hong Noh[2], and Do-Un Jeong[2(✉)]

[1] Department of Ubiquitous IT Engineering Graduate School, Dongseo University, Busan, Republic of Korea
[2] Division of Computer and Information Engineering, Dongseo University, Busan, Republic of Korea
dujeong@dongseo.ac.kr

Abstract. Modern society spends a lot of time sitting because of learning and work environments. However, sitting in the wrong posture increases musculoskeletal disorders, and affects student concentration in early childhood and adolescence, where musculoskeletal structures are formed. This paper developed a system to induce correct posture improve and seating habits for all ages. The proposed system measures the intensity, frequency, and distraction of posture changes. We tested 10 university students to evaluate the system performance and verified that the proposed system can effectively estimate posture classification and distraction.

Keywords: Posture correction · Distraction · Unconstrained

1 Introduction

Various systems assist managing your health, such as personal health care, smart health care, and disease management health care. Health management can be divided into disease prevention and disease treatment [1]. Modern medical services have shifted their focus from disease treatment to prevention, and a great deal of medical and IoT technology has been developed by the convergence of these medical services [2]. This study used IoT technology to implement a continuous health care monitoring system, in particular, monitoring posture and distraction in the sitting state during daily life. There are no current distraction development trends, so we defined research criteria based on and behavioral behavior of the family medicine department. Therefore, the proposed system employed in this study detects impulsive posture changes to estimate distraction.

2 Implementation of Distraction Estimation System

This paper implemented chair based distraction estimation during sitting, as shown in Fig. 1. The system incorporates three load cells below the seat in a regular triangle shape to measure body mass movement that occurs during posture changes, and estimates a

K.J. Kim et al. (eds.), *IT Convergence and Security 2017*,
Lecture Notes in Electrical Engineering 450,
DOI 10.1007/978-981-10-6454-8_4

distraction index using an attitude determination algorithm based on weight information and trigonometric centering. We developed PC and Android based monitoring systems using wired and wireless transmission. The proposed system is suitable for real time monitoring and continuous management and feedback.

Fig. 1. Proposed distraction monitoring system

2.1 Measurement

The measuring unit comprised three load cells (SB S-beam load cell, 100 kgf) arranged in an equilateral triangle to improve center estimate accuracy. The load cells generated signals with mV granularity based on the occupant's seating position. This signal was amplified using AD620, and AD620, which feature low offset voltage (50 uA), low input bias current (1 nA), and CMRR = 100 dB. Weight information was measured using a twin-T notch filter with 60 Hz cut-off frequency to eliminate power supply noise, and separated into right, left, and hip cell responses, then used to determine current posture and changes. Figure 2 shows posture change and weight shift for the implemented system, and Fig. 3 shows the placement and structure of the load cell.

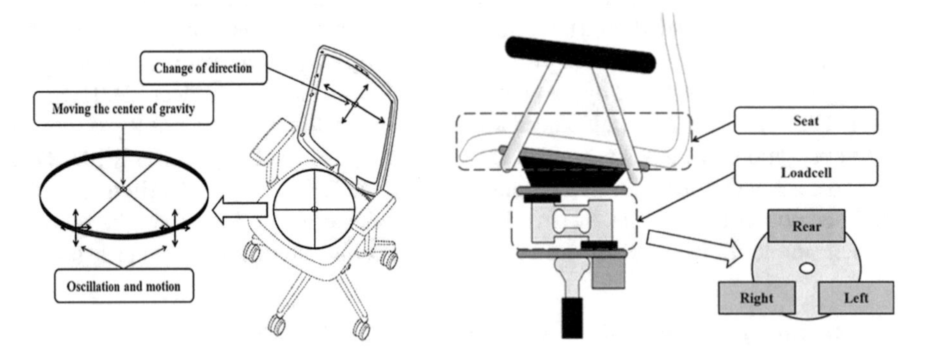

Fig. 2. Position of the change in posture change and weight shift

Fig. 3. Weight shift in the implementation system

2.2 Posture Change Estimate

We implemented an algorithm to estimate posture change intensity and frequency from real-time measurements using the trigonometric centering algorithm. This returned information of each load cell in coordinates, and estimates the center of mass movement direction and distance, as shown in Fig. 4.

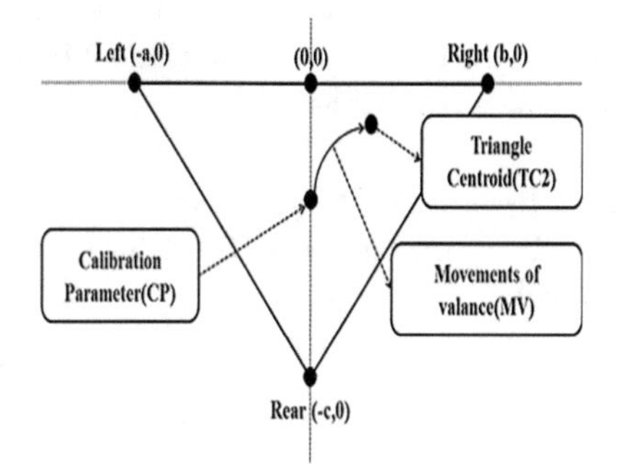

Fig. 4. Triangular center of gravity based motion algorithm

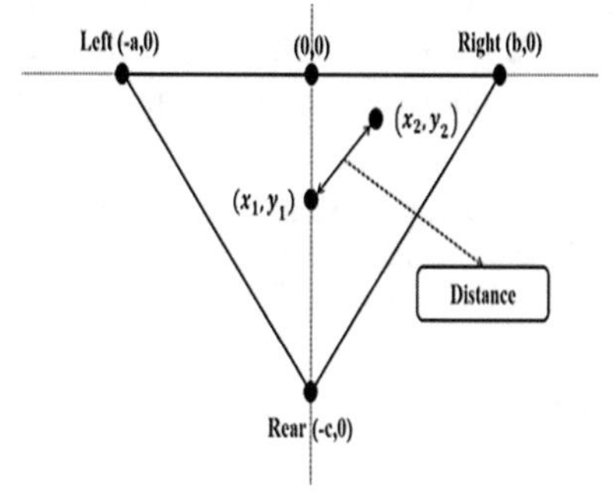

Fig. 5. Strength detection algorithm

3 Experiments and Results

To evaluate the proposed system's performance, we estimated distraction for 10 healthy college students while watching an audiovisual feature for 10 min. Table 1 shows that the subjects changed posture an average of 18 times during the feature, with average generation intensity = 2.70 and attitude detection = 92%. Thus, the proposed system was able to effectively estimate posture and distraction (from posture change) (Fig. 5).

Table 1. Posture and distraction data captured by the proposed system for 10 subjects

Subject	Proposed system summary statistics		
	Average intensity	Posture changes	Posture detection (%)
1	3	18	97
2	4	24	87
3	2	16	87
4	1	15	93
5	3	20	90
6	2	21	100
7	4	23	95
8	5	24	87
9	0	0	100
10	3	18	94
Avg	2.70	17.90	92.70

4 Conclusions

This study implemented distraction estimation from measured posture changes. The proposed system comprised a chair with three load cells to determine posture, and measured center of gravity movement intensity caused by posture change, which was converted to a distraction index. An experimental trial showed that the proposed system was able to discriminate nine common postures, and had high discrimination success.

Future research will analyze the success rate and the decline of the concentration of posture from long time scale measurements and video evaluation.

Acknowledgment. This research was supported by the Basic Science Research Program through the National Research Foundation of Korea (NRF) funded by the Ministry of Education (Grants 2015R1D1A1A01061131 and 2016R1D1A1B03934866).

References

1. Sok, Y.-Y., Kim, S.-H.: Integrated medical information system implementation for the u-healthcare service environment. J. Korea Contents Assoc. **14**(5), 1–7 (2014)
2. Harmin, M., et al.: Healthcare aide: towards a virtual assistant for doctors using pervasive middleware. In: Fourth Annual IEEE International Conference on PerCom Workshops 2006, pp. 6–495. IEEE (2006)
3. Lee, E.-Y., Seo, K.-E., Jung, W.-J., O'Sullivan, D.: Evaluation of the usefulness of smart good posture cushion. Korean J. Sports Sci. **25**(6), 1511–1521 (2016)

A Study of Job Competencies for Healthcare Social Work in Case-Based Discussion in Taiwan

Yi-Horng Lai[1(✉)], Hui-Yun Xiong[2], Shu-Chen Hsueh[2], and I-Jen Wang[2]

[1] Department of Healthcare Management, Oriental Institute of Technology, New Taipei City 22061, Taiwan
FL006@mail.oit.edu.tw

[2] Department of Social Work, Far Eastern Memorial Hospital, New Taipei City 22061, Taiwan
{sw19779,wij0211}@femh.org.tw, shu978@gmail.com

Abstract. Job competencies are central to every organization's human resource management system. Competencies can be used to help organizations create high performance, select and hire a workforce, and establish a foundation for training strategies. Social work, as a profession, has developed competency models for many specialized fields of social work. The purpose of this study was to identify the job competencies that exemplary healthcare social workers in case-based discussion in an effort to provide a foundation for a competency model. The participants selected for this study was exemplary performers of senior healthcare social workers. Deep interviews rendered data from which to analyze the opinions of five senior healthcare social workers in a medical center in Taiwan. A thematic analysis of the interview data revealed main competencies for healthcare social workers in case-based discussion are Ability to Decide Intervention Programs, Ability to Provide Individualized Care, and Cross Team Collaboration Ability. Results from this study have implications for healthcare organizations, social work education, and the professional development of healthcare social workers.

Keywords: Healthcare social work · Job competencies · Case-based discussion (CBD) · Analytic Hierarchy Process (AHP)

1 Introduction

Social workers are perceived to work as child welfare workers, individual and group counselors, and community organizers. It is important that employers recognize the value of social work skills in corporate and business settings. Some campus programs are now training social workers to lead social responsibility strategies, diversity initiatives, and employee wellness systems. Narrowing the gap between employer perception and reality requires social workers to adopt strategic marketing approaches that emphasize talent, competency, and knowledge base [1]. Therefore, social work skills and talents should be practiced in such a way that expands the definition and scope of social work and sustains the profession in a rapidly changing world society.

K.J. Kim et al. (eds.), *IT Convergence and Security 2017*,
Lecture Notes in Electrical Engineering 450,
DOI 10.1007/978-981-10-6454-8_5

Fields of social work include child welfare, gerontology, healthcare, mental and behavioral health, substance abuse, clinical social work, public welfare and community development, school social work, criminal justice, and international social work [2]. The competencies and skills of healthcare social workers are vital to the growth and improvement of the profession [3] because healthcare social workers permeates all areas of social work practice [4, 5]. Since case-based discussion competencies are unique to fields of healthcare social work practice [6] and healthcare is a prominent practice setting within the profession [7], a competency model is needed for healthcare social workers. To this date, research has not presented a competency model for healthcare social workers in case-based discussion, so this paper aims to make up this shortage.

Spencer and Spencer [8] indicated job competencies as characteristics that cause or predict behavior and performance as measured by a standard of minimally acceptable or superior performance. This means that there is evidence that indicates that the possession of the characteristics precedes and leads to effective performance. Competency methods give emphasis to what actually causes superior performance in a job [8]. There has not been any study for healthcare social workers in case-based discussion, the purpose of this study is to identify the job competencies that exemplary healthcare social workers in case-based discussion.

1.1 Job Competencies

A competency is a characteristic of an individual that causes or predicts behavior and performance as measured by a standard of minimally acceptable or superior performance [8]. Brownell and Goldsmith defined competencies as specific descriptions of the behaviors and personal characteristics that are required to be effective on the job [9]. Corporate human resource professionals generally define competency as an underlying characteristic of a person which results in effective performance on the job [10].

Competencies can be conceptualized into two categories: threshold competencies and differentiating competencies. Threshold competencies are exhibited by average performers. Differentiating competencies, on the other hand, are the behaviors that distinguish exemplary performers from average performers [8]. Therefore, the competence of an individual can be obtained by comparing the best instance of a performance with what is average [11].

1.2 Social Work in Healthcare

Healthcare was first introduced as a social work field of practice when Cannon and Cabot started the inclusion of social workers to physician teams to address the social problems that hindered patients' health in 1905 [1]. They found that the effectiveness of medical treatment was influenced by a patient's family state [12].

The goal of social work within healthcare settings is to contribute to high quality patient outcomes [13]. Further explained by Mayer [14] and Rosenberg and Weissman [15], the primary purpose of medical social work is to attend to the environmental and psychosocial problems affecting patients and their families. The National Association of Social Workers outlined general purposes of healthcare social work include to promote behaviors that contribute to the physical and emotional health of patients, to address psychosocial conditions that adversely affect patient health and well-being, to intervene to ensure patients receive services that maintain or improve their health and quality of life, and to help patients manage and adjust to their health conditions in order to realize maximum social functioning [16]. These objectives are satisfied through the implementation of diverse and complex practice functions.

In general, the main function of social work department in a hospital is discharge planning [13]. Discharge planning is focused on helping patients, families, and care-givers plan for the post-hospital care of patients [14]. Particularly, this function refers to the characteristic activities of assessment, case management, and information/referral management [13]. For cases in which patients are economically disadvantaged, financial assistance accompanies discharge planning activities in effort to help them find the means to pay for medical expenses [13]. For cases in which patients have psychosocial problems that impact the discharge plan, medical social workers carry out in-depth psychosocial evaluations and interventions ranging from supportive counseling to clinical social work services that include individual or group treatment [14].

2 Research Methodology

2.1 Research Framework

This study discussed the job competencies for healthcare social work in case-based discussion. This study employed the two main principles of core competencies as defined by senior healthcare social workers in a medical center in Taiwan (Far Eastern Memorial Hospital): Case Records Assessment and Clinical Ability Assessments. Regarding the sub-principles of Case Records Assessment and Clinical Ability Assessments, this study included those sub-principles proposed by senior healthcare social workers in a medical center in Taiwan. Finally, this study constructed the framework of indicators of core competencies for healthcare social work in case-based discussion.

2.2 Research Methods

The Analytic Hierarchy Process (AHP) was devised by Thomas Saaty [19]. AHP is an effective tool for dealing with complex decision making, and may help the decision maker to set priorities and make the decision. By reducing complex decisions to a series of pairwise comparisons, and then creating the results, the AHP helps to capture both subjective and objective features of a decision [20]. The AHP considers a set of evaluation criteria, and a set of alternative options among which the best decision is to be made.

The AHP makes a weight for each evaluation criterion according to the decision maker's pairwise comparisons of the criteria. The higher the weight, the more important the corresponding criterion. For a fixed criterion, the AHP assigns a score to each option according to the decision maker's pairwise comparisons of the options based on that criterion. The higher the score, the better the performance of the option with respect to the considered criterion. The AHP combines the criteria weights and the options scores, thus defining a global score for each option, and a consequent ranking. The global score for a given option is a weighted sum of the scores it obtained with respect to all the criteria.

The AHP is a very flexible and powerful tool because the scores, and therefore the final ranking, are got on the basis of the pairwise relative evaluations of both the criteria and the options provided by the user. The computations made by the AHP are always guided by the decision maker's experience, and the AHP can thus be considered as a tool that is able to translate the evaluations made by the decision maker into a multi-criteria ranking. The AHP is simple because there is no need of building a complex expert system with the decision maker's knowledge embedded in it.

For compute the weights for the different criteria, the AHP starts creating a pairwise comparison matrix A. The matrix A is a m × m real matrix, where m is the number of evaluation criteria considered. Each entry a_{jk} of the matrix A represents the importance of the jth criterion relative to the kth criterion. If $a_{jk} > 1$, then the jth criterion is more important than the kth criterion, while if $a_{jk} < 1$, then the jth criterion is less important than the kth criterion. If two criteria have the same importance, then the entry a_{jk} is 1. The entries a_{jk} and a_{kj} satisfy the following constraint as Eq. (1).

$$a_{jk} \times a_{kj} = 1 \tag{1}$$

Obviously, $a_{jj} = 1$ for all j. The relative importance between two criteria is measured according to a numerical scale from 1 to 9, where it is assumed that the jth criterion is equally or more important than the kth criterion. It is also possible to assign intermediate values which do not correspond to a precise interpretation. The values in the matrix A are by construction pairwise consistent as Eq. (1). On the other hand, the ratings may in general show slight inconsistencies. However these do not cause serious difficulties for the AHP.

After the matrix A is built, it is possible to derive from A the normalized pairwise comparison matrix A_{norm} by making equal to 1 the sum of the entries on each column, and each entry the average of a_{jk} of the matrix A_{norm} is computed as Eq. (2).

$$\bar{a}_{jk} = \frac{a_{jk}}{\sum_{l=1}^{m} a_{lk}} \tag{2}$$

After all, the criteria weight vector w is built by averaging the entries on each row of A_{norm}, as Eq. (3). This study calculated the AHP weights with Eq. (3) with R 3.4.0.

$$w_j = \frac{\sum_{l=1}^{m} \bar{a}_{jl}}{m} \tag{3}$$

2.3 Identifying Indicators of Core Competency

This study employed in-depth interviews coupled with a questionnaire on the importance of indicators to identify essential core competencies. Then, this study investigated the importance of the framework indicators and the appropriateness of the classifications. The results indicate that the research subjects agreed with the classification of the principles. The indicator framework of the core competencies is presented in Fig. 1.

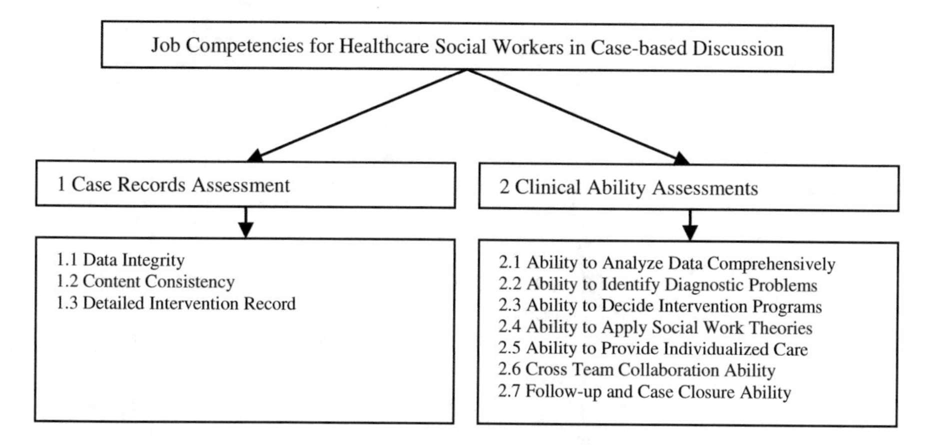

Fig. 1. Modified indicator framework of the core competencies of healthcare social workers in case-based discussion.

3 Results

After verifying the essential indicator framework of the core competencies, this study asked the panel experts to compare the importance of any two principles via questionnaires, and adopted the AHP to calculate the weight of each principle. The expert panel was comprised of senior healthcare social workers. Deep interviews rendered data from which to analyze the opinions of five senior healthcare social workers in medical centers.

The result of the AHP is shown in Table 1. The consistency ratio (CR) of both the main principles and sub-principles was less than 0.1, reaching a level of consistency. Regarding the main principles, Clinical Ability Assessments (weight = 0.824) has the highest weightage, followed by Case Records Assessment (weight = 0.176), indicating that senior healthcare social workers mostly focus on Clinical Ability Assessments in case-based discussion in Taiwan. Regarding the sub-principles of Clinical Ability Assessments, Ability to Decide Intervention Programs (weight = 0.265) has the highest weightage, followed by Ability to Provide Individualized Care (weight = 0.211) and Cross Team Collaboration Ability (weight = 0.171). In the sub-principles of Case Records Assessment, Detailed Intervention Record (weight = 0.496) has highest weightage, followed by Data Integrity (weight = 0.345) and Content Consistency (weight = 0.159).

Based on the weights of the main and sub-principles, this study inferred that the most essential core competency for healthcare social workers is the Ability to Decide Intervention Programs (total weight = 0.219), followed by Ability to Provide Individualized Care (total weight = 0.174), and Cross Team Collaboration Ability (total weight = 0.141).

Table 1. Weights of the core competencies of healthcare social workers in case-based discussion.

Main criteria	Weight	Sub-criteria	Weight	Total weight
1 Case Records Assessment[a]	0.176	1.1 Data Integrity	0.345	0.061
		1.2 Content Consistency	0.159	0.028
		1.3 Detailed Intervention Record	0.496	0.087
2 Clinical Ability Assessments[b]	0.824	2.1 Ability to Analyze Data Comprehensively	0.142	0.117
		2.2 Ability to Identify Diagnostic Problems	0.169	0.139
		2.3 Ability to Decide Intervention Programs	0.265	0.219
		2.4 Ability to Apply Social Work Theories	0.091	0.075
		2.5 Ability to Provide Individualized care	0.211	0.174
		2.6 Cross Team Collaboration Ability	0.171	0.141
		2.7 Follow-up and Case Closure Ability	0.093	0.077

[a]The CI and CR for the sub-criteria of Case Records Assessment are 0.04 and 0.07.
[b]The CI and CR for the sub-criteria of Clinical Ability Assessments are 0.03 and 0.02.

4 Conclusions and Suggestions

4.1 Conclusions

The present study sought to fill a vacuum in the studies exploring healthcare social work as existing bodies of knowledge have failed to present a competency model for healthcare social workers. A comprehensive review of the study cited felicitous subject matters and revealed that much of the research on healthcare social work is focused on the management tasks and activities of the hospital social workers. Therefore, this investigation was the first attempt to explore and define job competencies for the healthcare social workers in case-based discussion in Taiwan.

This study constructed a recruitment framework of the core competencies of healthcare social workers in case-based discussion through in-depth interviews. This study concluded the core competencies of the core competencies of healthcare social workers in case-based discussion are two main principles (Case Records Assessment and Clinical Ability Assessments) and 10 sub-principles. Of the principles, Ability to Decide

Intervention Programs, Ability to Provide Individualized Care, and Cross Team Collaboration Ability had the highest weightage.

4.2 Suggestions

The results of this study suggest that the job competencies of healthcare social workers in case-based discussion are products of a complex set of behaviors and characteristics demonstrated by exemplars in the field. Given the case-based discussion required of healthcare workers, it is paramount that social work practice in healthcare organizations be predicated on job competencies. Such competencies may be an important topic in healthcare workers training. It is critical that healthcare organizations, professional associations, scholars, and researchers alike extend their understanding of job competencies beyond what has been so commonly referred to as skills. These research findings can be used as a foundation to build a competency model for healthcare social workers.

Based on the result of this study, in addition to the enrichment of healthcare social workers, graduates from departments related to social works should regard Ability to Decide Intervention Programs as essential to successfully entering the job market. In case-based discussion, this study suggests that students polish their Ability to Decide Intervention Programs by participating in extracurricular activities.

The scope of this research focused on the healthcare social work in Taiwan. This study suggests that future research investigate the recruitment principles of the healthcare social workers in different countries and analyze regional differences.

References

1. Spitzer, W., Silverman, E., Allen, K.: From organizational awareness to organizational competency in health care social work: the importance of formulating a profession-in-environment fit. Soc. Work Health Care **54**(3), 193–211 (2015)
2. National Association of Social Workers: Practice (2013). http://www.naswdc.org/practice/default.asp. Accessed 1 May 2017
3. Perlmutter, F.D.: Ensuring social work administration. Adm. Soc. Work **30**(2), 3–9 (2006)
4. Ginsberg, L.H.: Careers in Social Work, 2nd edn. Allyn & Bacon, Needham Heights (2001)
5. Menefee, D.T., Thompson, J.J.: Identifying and comparing competencies for social work management. Adm. Soc. Work **18**(3), 1–25 (1994)
6. Preston, M.S.: The direct effects of field of practice on core managerial role competencies: a study across three types of public sector human service agencies. Adm. Soc. Work **32**(3), 63–83 (2008)
7. Whitaker, T., Weismiller, T., Clark, E.: Assuring the Sufficiency of a Frontline Workforce: A National Study of Licensed Social Workers. Executive Summary. National Association of Social Workers, Washington, D.C. (2006)
8. Spencer, L.M., Spencer, S.M.: Competence at Work: Models for Superior Performance. Wiley, New York (1993)
9. Brownell, J., Goldsmith, M.: Commentary on "meeting the competency needs of global leaders: a partnership approach": an executive coach's perspective. Hum. Resour. Manag. **45**(3), 309–336 (2006)

10. Lucia, A., Lepsinger, R.: The Art and Science of Competency Models: Pinpointing Critical Success Factors in Organizations. Jossey-Bass/Pfeiffer, San Francisco (1999)
11. Gilbert, T.F.: Human Competence: Engineering Worthy Performance. Pfeiffer, San Francisco (2007). (Tribute Edition)
12. Dhooper, S.S.: Social Work in Health Care in the 21st Century. SAGE Publications, Inc., Thousand Oaks (1997)
13. Bixby, N.B.: Crisis or opportunity: a healthcare social work director's response to change. Soc. Work Health Care **20**(4), 3–20 (1995)
14. Mayer, J.B.: The effective healthcare social work director: managing the social work department at Beth Israel Hospital. Soc. Work Health Care **20**(4), 61–72 (1995)
15. Rosenberg, G., Weissman, A.: Preliminary thoughts on sustaining central social work departments. Soc. Work Health Care **20**(4), 111–116 (1995)
16. National Association of Social Workers Center for Workforce Studies: Social work salaries by race & ethnicity: occupational profile (2011). http://workforce.socialworkers.org/studies/profiles/Race%20and%20Ethnicity.pdf. Accessed 1 May 2017
17. Ruster, P.L.: The evolution of social work in a community hospital. Soc. Work Health Care **20**(4), 73–88 (1995)
18. Burns, L.R., Bradley, E.H., Weiner, B.J.: Shortell and Kaluzny's Health Care Management Organization Design and Behavior, 6th edn. Delmar, Clifton Park (2012)
19. Saaty, T.L.: How to make a decision: the analytic hierarchy process. Eur. J. Oper. Res. **48**(1), 9–26 (1990)
20. Calantone, R.A., Di Benedetto, R.A., Meloche, M.: The analytic hierarchy process as a technique for retail store location selection. J. Bus. Strateg. **6**(1), 61–74 (1989)

Observation of Continuous Blood Pressure with Posture Change Using a Wearable PTT Monitoring System

Yun-Hong Noh[1], Ji-Yun Seo[2], and Do-Un Jeong[1(✉)]

[1] Department of Ubiquitous IT Engineering Graduate School, Dongseo University, Busan, Republic of Korea
dujeong@dongseo.ac.kr

[2] Division of Computer and Information Engineering, Dongseo University, Busan, Republic of Korea

Abstract. Smart healthcare services are continuously is increasing due to rapid industrialization and increasing chronic diseases. Significant research has focused on monitoring the health of an individual health in real time and to prevent disease in advance. This paper implemented a wearable ECG and PPG measurement system to conveniently monitor personal health in daily life. The proposed system used pulse transit time, the time difference between ECG and PPG R-peak signals. To evaluate the proposed system performance, we compared it with an existing commercial system to observe blood pressure and pulse wave changes according to posture, and validated the effectiveness of the proposed system.

Keywords: ECG (Electrocardiogram) · PPG (Photoplethysmogram) · PPT (pulse transit time) · Smart healthcare · Real-time monitoring

1 Introduction

Chronic diseases are increasing due to rapid industrialization, with consequential strong and increasing interest in smart healthcare. The convergence of information communication technologies and services has removed time and space restrictions. U-healthcare has conducted a number of studies to noninvasively measure ECG and PPG, which contain health information among the vital signs. ECG in particular provides has the most basic information to check health status, and ECG dislocation changes have an almost one-to-one correlation with blood pressure and blood flow changes. Thus, the ECG probes heart function [1]. PPG measures the pulse wave transmitted to the peripheral nerve. When there is an abnormality or change in the body, the pulse wave shape changes and propagation speed increases. Previous studies have also investigated arterial perfusion and blood vessel tension by analyzing pulse transit time (PTT), which is closely related to the blood vessel quality, by simultaneously measuring PPG and ECG [2]. This paper implemented an ECG and PPG measurement system that can be worn on the body and provide convenient health monitoring in daily life. The proposed system calculated PTT as the time difference between PPG and ECG R-peak signals. Thus, we observed blood pressure and pulse wave transmission time changes according to posture.

K.J. Kim et al. (eds.), *IT Convergence and Security 2017*,
Lecture Notes in Electrical Engineering 450,
DOI 10.1007/978-981-10-6454-8_6

2 Posture Change and Pulse Transit Time Measurement System

PTT is the time between ECG and PPG R-peaks, measured on peripheral parts of the human body. PTT properties dependent on blood vessel quality, distance, diameter, and structural of the blood vessel wall [3]. Thus, PTT represents physical characteristics of the not the local vessel sites blood vessels from the heart to the specific site, as compared to the pulse wave transit velocity. With the advent of new and simpler measurement systems, PTT utilization is rapidly increasing.

To measure the ECG and PPG, a sensor unit was constructed that included an Ag-AgCl surface electrode and reflection type PPG sensor, analog signal processing device for detecting and amplifying ECG and PPG signals, wireless sensor node for converting analog signals to digital signals and transmitting data wirelessly, and a monitoring program for displaying and storing data on a PC (Figs. 1 and 2).

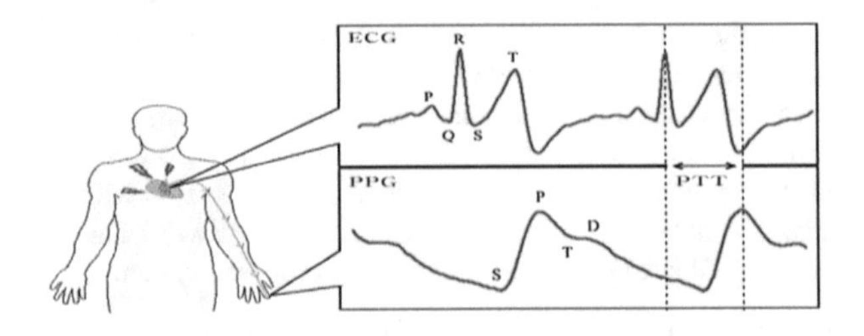

Fig. 1. Pulse transit time concept

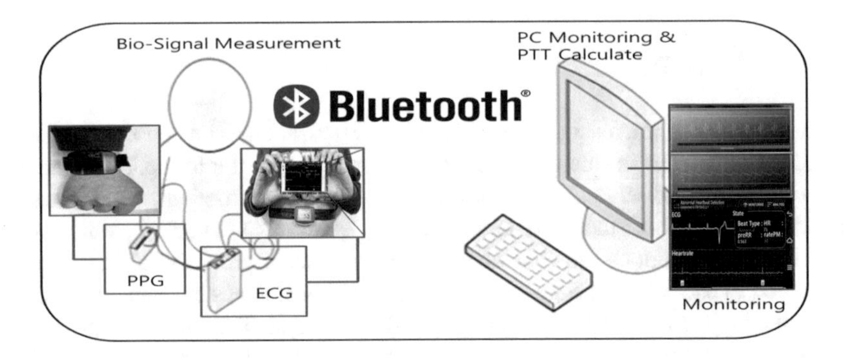

Fig. 2. ECG and PPG measurement system

3 Experiment and Results

3.1 Comparison with Commercial Products

This study compared the proposed bio-signal system with the P400 PPT system (Physiolab). Ten healthy college students were selected and data measured for 5 min, as shown in Fig. 3. The proposed PTT system provides very similar profiles to the commercial P400 system, although the absolute values differ somewhat.

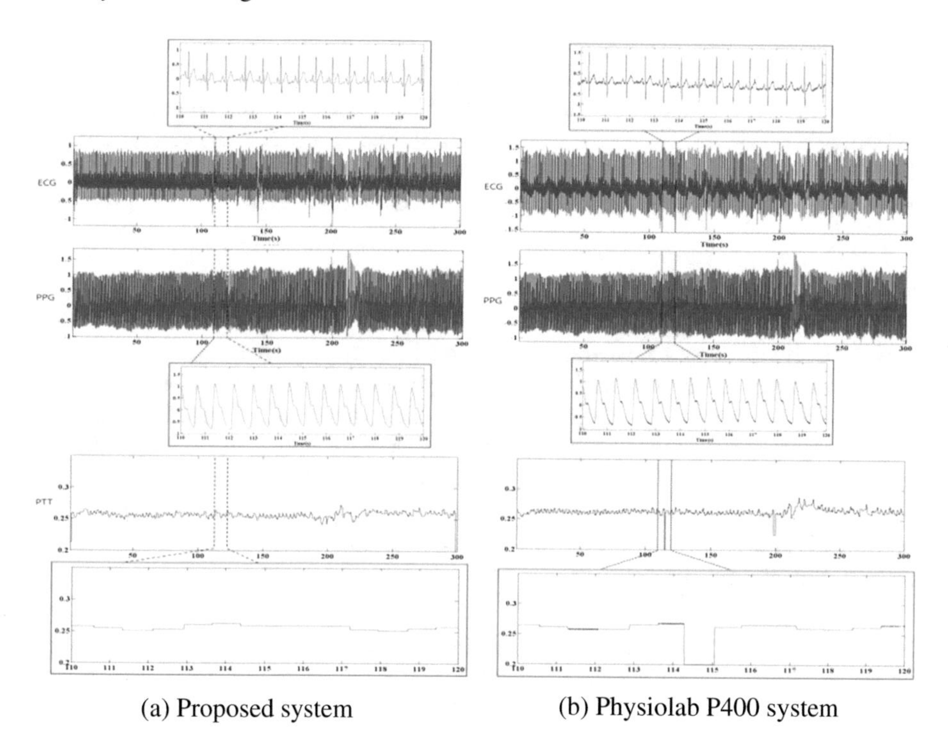

(a) Proposed system (b) Physiolab P400 system

Fig. 3. Pulse transit time measured by different systems

Table 1. Pulse transit time (PTT) variation with posture [ms]

	PTT sitting (ms)		PTT lying (ms)	
	Mean	SD	Mean	SD
Object 1	243.30	4.90	246.10	5.10
Object 2	241.79	5.17	245.10	7.20
Object 3	231.25	4.59	237.53	5.70
Object 4	235.67	5.23	240.63	9.17
Object 5	243.53	4.86	253.24	6.71
Object 6	237.49	4.92	242.65	6.14
Object 7	240.96	11.21	246.440	10.69
Object 8	239.60	5.34	246.64	6.35
Object 9	208.49	3.91	213.13	4.14
Object 10	231.09	4.66	238.55	5.53

Note: SD = sample standard deviation

3.2 Pulse Transit Time Change with Posture Change

Blood pressure and blood vessel status differ with body position. Previous studies have shown that PTT decreases for increasing blood pressure and vice versa. The current study used ECG and PPG to observe PTT and blood pressure changes due to posture change. Ten healthy volunteers had ECG and PPG measured simultaneously in lying and sitting position for 5 min, as shown in Table 1. Subjects sitting showed shorter PTT than lying, because the sitting posture allows the arm measurement point to fall below the heart, which increases hydrostatic pressure in the blood vessel.

4 Conclusions

This paper implemented a wearable ECG and PPG measurement system to provide convenient health monitoring in daily life. The performance of the proposed system was compared to an existing commercial system, which validated the effectiveness of the proposed system and confirmed measurable PTT changes with posture change. Thus, it was possible to monitor blood pressure change through PTT, which could be used for noninvasive and continuous cardiovascular health monitoring. Future studies will estimate blood pressure changes and disease status from PTT changes, compared with various clinical instruments.

Acknowledgment. This research was supported by the Basic Science Research Program through the National Research Foundation of Korea (NRF) funded by the Ministry of Education (Grants 2015R1D1A1A01061131 and 2016R1D1A1B03934866).

References

1. Ezzati, M., Lopez, A.D., Rodgers, A., Murray, C.J.L.: Comparative Quantification of Health Risks: Global and Regional Burden of Disease Attributable to Selected Major Risk Factors. World Health Organization, Geneva (2004)
2. Lee, C.-L., Kim, K.-H.: A preliminary study on continuous measurement blood pressure using PTT. KIEE **10**, 380–381 (2013)
3. Poon, C.C.Y., Zhang, Y.T.: Cuff-less and noninvasive measurements of arterial blood pressure by pulse transit time. In: Proceedings of the 27th Annual Conference of IEEE-EMBC, pp. 5877–5880 (2005)

Real-Time ECG Monitoring System Based on the Internet of Things Using an Optimum Transmission Technique

Yun-Hong Noh[1], Ji-Yun Seo[2], and Do-Un Jeong[1(✉)]

[1] Department of Ubiquitous IT Engineering Graduate School, Dongseo University, Busan, Republic of Korea
dujeong@dongseo.ac.kr

[2] Division of Computer and Information Engineering, Dongseo University, Busan, Republic of Korea

Abstract. This paper implemented an efficient monitoring system using the cloud based processing and storage capacity. The proposed system compressed electrocardiogram data and transmitted it to the cloud to enable more efficient processing than could be achieved from the limited resources of the measurement system. We confirmed the proposed method's suitability using the MIT/BIH database, and showed it provided good compression ratio, high recovery success rate, and low computational complexity.

Keywords: ECG compression · Template matching · IoT cloud based

1 Introduction

ECG (Electrocardiogram) signals reflect heart activity, and healthy hearts show regular period and/or rhythm. Analyzing ECG signal can identify the presence or absence of lesion site(s) [1], and can be used to diagnose cardiovascular disease by estimating heart and physiological abnormalities. Previous research has used real-timc classification to provide user centered services, but these require significant calculation overhead for accurate analysis and are not suitable for continuous monitoring [2]. Therefore, this paper implemented a wearable system to monitor ECGs in real time and daily life, compressing the data and transmitting it to the cloud for analysis. Thus, the proposed system implemented an IoT based healthcare system to detect abnormal electrocardiograms and notify the user only if abnormalities were identified to minimize data transfer (Fig. 1).

2 Cloud Based Data Compression and Transmission

We implemented a template matching electrocardiogram compression algorithm that maintained compression ratio and complexity without requiring a complicated calculation process, as shown in Fig. 2. The template matching method first detects the ECG R-peak, is a characteristic ECG point, and then generates a template, and calculates the

K.J. Kim et al. (eds.), *IT Convergence and Security 2017*,
Lecture Notes in Electrical Engineering 450,
DOI 10.1007/978-981-10-6454-8_7

similarity between the template and the actual ECG signal. Only this latter information (and the template ID) is saved.

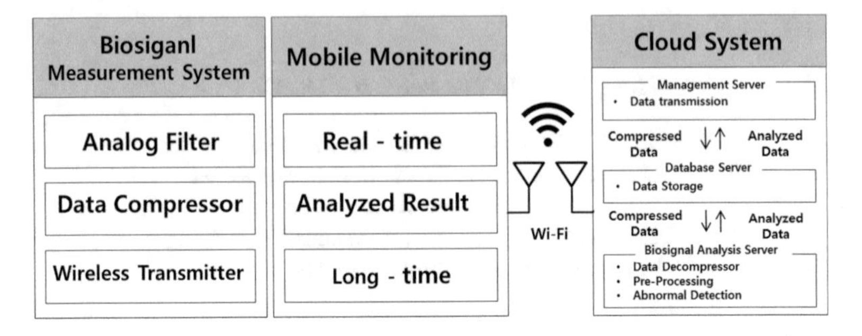

Fig. 1. Proposed real-time electrocardiogram monitoring system

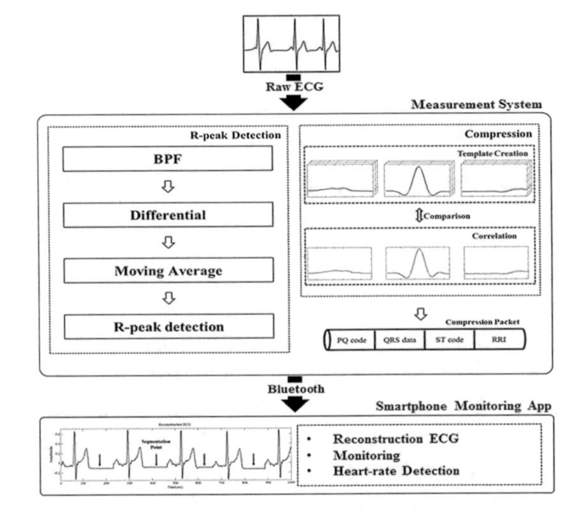

Fig. 2. Proposed electrocardiogram compression algorithm

3 Real-Time Monitoring Using the Internet of Things

Cloud based ECG analysis was implemented using a wireless monitoring system comprising Screen View, Communication, Pattern Data, and Data Management classes. The Data Manager distinguishes wirelessly transmitted data packets and stores them in each of the connection list objects. The Pattern Data Manager loads the indicated ECG normal heartbeat template and the actual ECG data from the transmitted data packet, and passes this to the Monitoring Graph Section for real-time monitoring (Fig. 3).

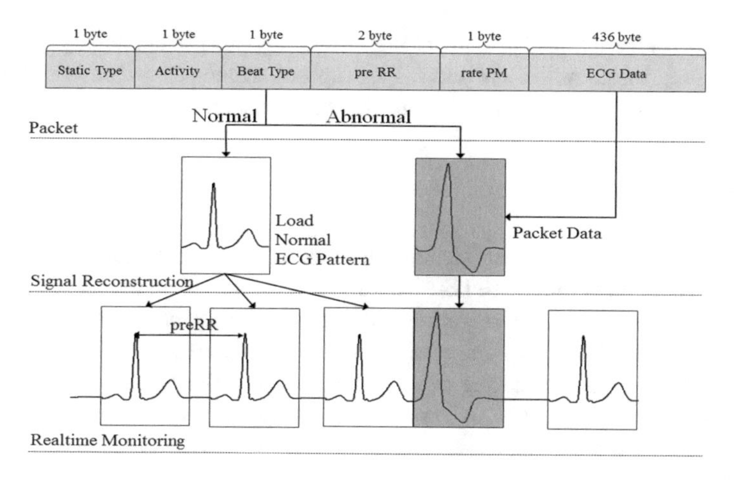

Fig. 3. Cloud based signal restoration and real-time monitoring

4 Experiments and Results

We used the MIT-BIH Normal Sinus Rhythm database to evaluate the proposed algorithm. Data were extracted from 10 records for 10 min. We employed the compression ratio (CR), percent root mean square difference (PRD), and peak detection accuracy (PDA) as performance metrics, as shown in Table 1. Average CR = 7.94 and average PRD = 5.33, which was significantly lower <than what>. However, the compression rate increased with increasing amount of information required to express the measured cycle, and the reconstruction rate can be reduced due to the signal high frequency component that

Table 1. Compression results using the <full phrase for MIT-BIH here> (MIT-BIH) Normal Sinus Rhythm database

MIT-BIH	CR	PRD	PDA (%)
16265m	6.64	4.04	100
16483m	7.81	4.50	
16539m	8.04	5.70	
16773m	8.76	5.05	
16786m	9.47	3.63	
16795m	9.75	5.91	
17052m	8.11	5.45	
17453m	8.52	4.77	
18177m	6.43	6.82	
19088m	5.88	7.49	
Average	7.94	5.33	

Notes: CR = compression ratio, PRD = percent root mean square difference, (PRD), PDA = peak detection accuracy

can occur at low sampling. Thus, lower high frequency component noise produced more favorable compression ratios. Therefore, the proposed method can achieve high compression ratio with high sampling rate in a miniaturized measurement instrument.

5 Conclusions

This study implemented a wearable system to monitor ECGs in real-time in daily life. The proposed system compressed ECG data and transmitted it to the cloud for analysis. We implemented an ECG classification and feature point detection method in the measurement device to efficiently compress ECGs with the limited available resources. The proposed compression technique was evaluated using the MIT/BIH Database, and showed goo compression ratio, high restoration success, and low computation overhead. Thus, the proposed a technique is suitable for deployment as effective real-time ECG monitoring.

Acknowledgments. This research was supported by the Basic Science Research Program through the National Research Foundation of Korea (NRF) funded by the Ministry of Education (Grants 2015R1D1A1A01061131 and 2016R1D1A1B03934866).

References

1. Shin, S.-C., Kang, J.-H., Kim, S.-H.: Detection of ECG characteristic points for heart disease diagnosis. KIISE **29**(2), 199–201 (2002)
2. Kim, H., et al.: A configurable and low-power mixed signal SoC for portable ECG monitoring applications. IEEE Trans. Biomed. Circuits Syst. **8**(2), 257–267 (2014)

A Study on Programs Applying the Internet of Things (IoT) for Prevention of Falls in the Elderly

SeungAe Kang(✉)

Department of Sport and Healthcare, Namseoul University, Cheonan, Korea
sahome@nsu.ac.kr

Abstract. In this study, we presented the trends of IoT (Internet of Things) used to prevent, predict and detect falls in advance and the measures that could help spur application of IoT-based fall prevention technologies. The application of technology for fall prediction and detection can be divided at large into the image data analysis method and sensor data analysis method which focuses on data from sensors detecting physical activities. The applications of sensors include acceleration sensors, air pressure sensors, vibration sensors, gyroscopes, audio sensors, etc. The IoT expected to be further expanded in conjunction with big data analysis and home care service in the period ahead. For that, simplification of easy-to-operate device interfaces, linkage with home care services, more secure information collection and management should be ensured through synchronization with various devices and more rigorous security control of collected information.

Keywords: IoT · The elderly · Falls prevention · Sensor · Device · Home care

1 Introduction

The senescent population is confronted with various physical and psychological challenges. The degradation of physical functions, caused by aging, increases the risks of prevalence of many different diseases, accompanied by psychological contraction, which leads to an increase in costs incurred form diseases related to aging. Furthermore, aging causes muscles and bones to be weakened, leading to an unforeseen fall accidents that hinder everyday life of the elderly. After experience with fall accident, the decline in recovery ability and psychological fear make the aftereffects persistent. According to the "Burden of Diseases on Koreans" published in 2016, fall accident topped the list of diseases, even outstripping the cancer. One-third of the population aged 65 and older experience fall at least once a year [1]. Fractures caused by falls can increase the risk of complications in the elderly due to prolonged recovery time. Furthermore, the dysfunction and pain caused by fall accidents erode the quality of life and lead to higher mortality rate. To ensure safe and independent lives of those at senescence, it would be very important to prevent fall accidents that have recently increased.

With the advancement of IT (Information Technology), state-of-art electronic devices and industry are developing rapidly. In addition, technologies that can underpin protection of the elderly have been thrust into limelight as Korea is making a transition

K.J. Kim et al. (eds.), *IT Convergence and Security 2017*,
Lecture Notes in Electrical Engineering 450,
DOI 10.1007/978-981-10-6454-8_8

to the aged society where life expectancy is expected to rise to 100 years. Some robots of medical applications have been developed to assist patients with dementia, and the need for technology for care for the elderly has been ever more increasing.

This study was intended to examine the technologies applying the IoT (Internet of Things) to prevent fall accidents in the elderly and looked into related cases as part of efforts to mitigate the risks of fall in senescent population and to present efficient countermeasures that could reduce the risks of fall experienced by the elderly.

2 Technologies Applying the IoT for Prevention and Detection of Falls

The Internet of Things (IoT) is defined as a technology that provide users with services of optimal useful value by connecting the network of intelligent devices, capable of determining the situations and learning, to a gigantic net such as internet, thus bundling them into a single frame [2, 3]. Recently, IoT has expanded beyond the boundaries of Bluetooth and near-field communication (NFC) into IoT platform using the smartphones and found its applications even in healthcare field.

The elderly who experienced falls see a sharp decline in their own physical activities due to fear of falling, which leads to weakened sense of balance and musculoskeletal

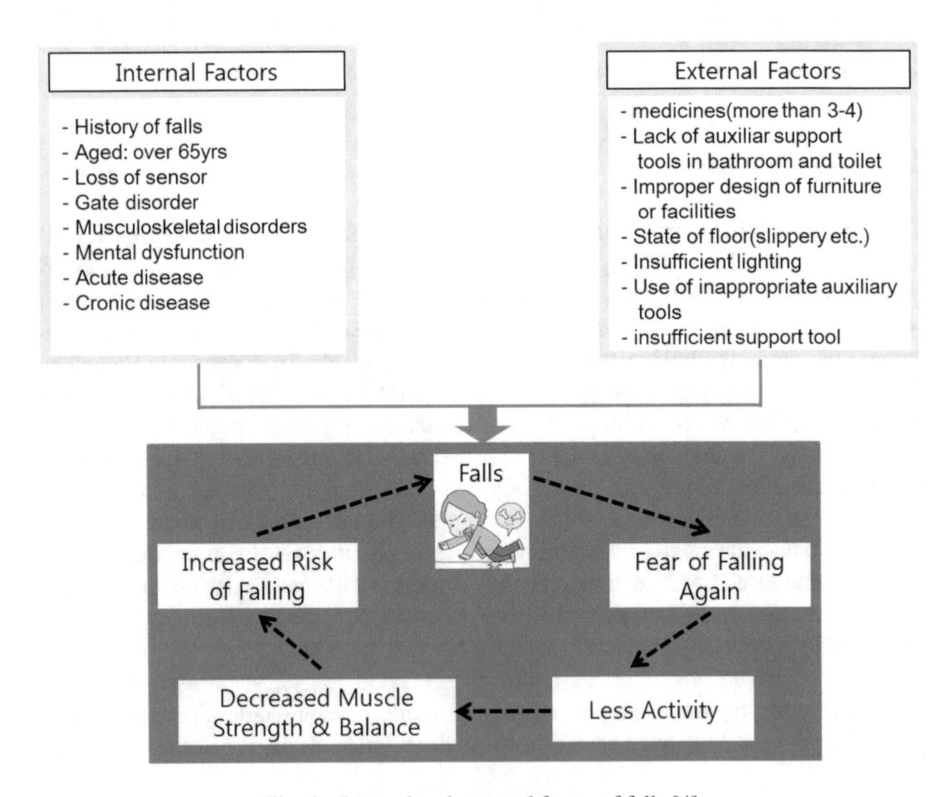

Fig. 1. Internal and external factor of falls [4]

loss and increases the risks of repeated fall accidents. Therefore, prevention of falls is important to avoid repetitive experience of falls. Risk factors for fall accident are classified into internal factors and external factors, as shown in Fig. 1.

If gait patterns and walking disorders are detected and analyzed by using the IoT technology among those factors, falls can be predicted, prevented, and detected. Amid the increase in senescent population and rising needs for better quality of life, IoT technology application is considered to represent an area with great growth potential in the field of elderly care in the period ahead. The application of technology for fall prediction and detection can be divided at large into the image data analysis method and sensor data analysis method which focuses on data from sensors detecting physical activities [3]. The image data analysis method involves tracking the behaviors of targets recognized by camera, which poses the risks of infringement upon privacy. Furthermore, equipment problems should be considered when target's radius of action is large while the importance of network reliability should be taken into consideration when image data are transferred. For sensor data analysis, the problem of image data analysis can be mitigated because the sensors analyze the motions of target. Examples of applications of sensors include acceleration sensors, air pressure sensors, vibration sensors, gyroscopes, audio sensors, etc.

Studies that presented the fall detection algorithms involving the use of wearable acceleration sensors showed that the methods of wearing or attaching the acceleration sensor were different. The sense of foreign material or difficulty with sensor attachment, etc., were reported, depending on the attachment position of sensors such as acceleration sensor attached to the waist [5], acceleration sensor attached to mobile phones [6], sensors attached to the chest [7], sensors attached to the ear [8], and sensors attached to the wrist [9]. In addition, sensor used in game machines can be utilized to check the outline of object through infrared rays. By using this method, the gait of the elderly can be analyzed and fall accident can be predicted 3 months in advance. Air pressure sensor and vibration sensor can be used to detect fall accidents [10]. Besides, real-time fall detection system [11] using the acceleration sensor and tilt sensor was developed, along with smart cane [12] designed to prevent falls using motion sensors and GPS, including the fall detection system using the data of angular speed variation in gyroscope [13] or utilizing the analytical data obtained from acceleration sensors and audio sensors or utilizing the data from built-in acceleration sensor of smartphone [14]. Moreover, studies [15] have been conducted to predict the fall based on detection and analysis of the elderly behaviors through convergence of IoT and big data technologies.

3 Measures for Stimulation of Fall Prevention Technologies

To stimulate the application of IoT-based fall prevention technologies, the following measures are required:

- **Simplification of easy-to-operate device interfaces:** In order to prevent and detect the fall of the elderly in advance, it is important to ensure the interface design easily accessible to the elderly who are the users of the device. Elderly-friendly interface design needs to be developed which considers the use of color with good discrimination which can complement the degraded eyesight and cognitive functions of the elderly, simple structure of menu layout that can be distinguished at a glance, and use of fonts [16] with high legibility and high information deliverability.

- **Linkage with home care services:** More specific fall prediction and prevention would be possible if various information related to the fall is collected and analyzed based on frequent monitoring of health conditions and daily activities of the elderly through cloud-based home care services synchronized with smart home (based on the IoT). This will enable interconnection among home care for the elderly, treatment and management of their chronic diseases, prevention of fall, etc., thus reducing medical costs. Various services will be able to be delivered in conjunction with visiting care and elderly care business in the period ahead.
- **Security Reinforcement:** Personal health information measured and collected through home care services, smart phones, wearable devices, etc., requires strict security maintenance and control. It is considered urgent to resolve the issues of security, given that the problem of personal information leakage has constantly occurred and controversy has swirled over such disclosure of personal data due to transmission of data via wired and wireless networks. The linkage with home care service is expected to expand in the future. For that, resolution of security problems is urgent with respect to various elements related to IoT such as network, platform, devices, sensors, etc. [13]. The need for more rigorous information security and control systems should be further increased, along with the need for clarification of responsibilities among stakeholders in the event of information leakage.

4 Conclusion

The rapid increase in elderly population aged 65 and older and their increased experience with falls are directly associated with health problems and quality of life of the elderly. In this study, we presented the trends of IoT used to prevent, predict and detect falls in advance and the measures that could help spur application of IoT-based fall prevention technologies. The IoT, which includes sensor network, has been replaced through sensor data analysis that complements conventional image data analysis and is expected to be further expanded in conjunction with big data analysis and home care service in the period ahead. For that, more secure information collection and management should be ensured through synchronization with various devices and more rigorous security control of collected information.

References

1. Korea insurance news: Falls is more severe than cancer, 19 March 2017. Press
2. Sin, D.H., Jeong, J.Y., Kang, S.H.: Internet of Things trend and vision. Rev. Korean Soc. Internet Inf. **14**(12), 32–46 (2013)
3. Jeong, P.S., Cho, Y.H.: Fall detection system based Internet of Things. J. Korea Inst. Inf. Commun. Eng. **19**(11), 2546–2553 (2015)

4. The Joint Commission: Comprehensive Accreditation Manual for Hospitals: The Official Handbook. The Joint Commission on Accreditation of Healthcare Organizations, Oakbrook Terrace (2006)
5. Zhang, T., Wang, J., Xu, L., Liu, P.: Fall detection by wearable sensor and one-class SVM algorithm. Lecture Notes in Control and Information Sciences, pp. 858–863 (2006)
6. Zhang, T., Wang, J., Liu, P., Hou, J.: Fall detection by embedding in accelerometer in cellphone and using KDF algorithm. IJCSNS Int. J. Comput. Sci. Netw. Secur. **6**(10), 277–284 (2006)
7. Hwang, J.Y., Kang, J.M., Jang, Y.W., Kim, H.C.: Development of novel algorithm and real-time monitoring ambulatory system using bluetooth module for fall detection in the elderly. In: Proceedings of 26th Annual International Conference on IEEE EMBS, pp. 2204–2207 (2004)
8. Lindermann, U., Hock, A., Stuber, M., Keck, W., Beeker, C.: Evaluation of a fall detector based on accelerometers: a pilot study. Med. Biol. Eng. Comput. **43**(5), 548–551 (2005)
9. Kim, N.S.: An efficient methodology of fall detection for ubiquitous healthcare. J Korean Inst. Inf. Technol. **8**(8), 133–140 (2010)
10. Chosun daily news: Falls in the elderly is more severe than cancer, 22 February 2017. Chosun.com. Press
11. Kim, S.H., Park, J., Kim, D.W., Kim, N.G.: The study of realtime fall detection system with accelerometer and tilt sensor. J. Korean Soc. Precis. Eng. **28**(11), 1330–1338 (2011)
12. Kim, T.K., Ro, C.W., Yoon, J.W.: Development of smart stick using motion sensing and GPS for elderly users' safety. J. Korea Converg. Soc. **7**(4), 45–50 (2016)
13. Bourke, A.K., Lyons, G.M.: A threshold-based fall-detection algorithm using a bi-axial gyroscope sensor. Med. Eng. Phys. **30**, 84–90 (2008)
14. Lee, G.E., Lee, J.W.: Comparison study of web application development environments in smartphone. J. KOCON **10**(12), 155–163 (2010)
15. Seoul National University Bundang Hospital Signs MOU for fall prevention project with SAP-SpoMedex. http://www.docdocdoc.co.kr/newspress.2016.07.21. Press
16. Huh, W.W., Cho, J.K.: App UI design research for improving the usability of the silver generation. Asia-Pacific J. Multimedia Serv. Converg. Art Hum. Sociol. **6**(10), 565–572 (2016)

Smart Care to Improve Health Care for the Elderly

SunYoung Kang[1] and SeungAe Kang[2(✉)]

[1] Department of Physical Education, Korea University, Seoul, Korea
1010kang@hanmail.net

[2] Department of Sport and Healthcare, Namseoul University, Cheonan, Korea
sahome@nsu.ac.kr

Abstract. Smart care technology which has made strides is recognized vital in the aging society. This study was intended to examine trends and outlook of smart care by type of elderly and situation to spur improvement in health management for the elderly. Smart cares by type of elderly and situation were elderly-friendly smart home, smart care for management of chronic diseases in the elderly, and smart care for the elderly living alone. Smart care is needed to be implemented actively not only to reduce medical costs and improve the quality of medical services but also to ensure safety and quality of life for the elderly. However, multi-faceted efforts need to be made to help overcome the difficulties arising from degraded cognitive function of the elderly, considering the basic orientation of smart care such as mobile, smart, cloud computing, IoT, etc. For that, it would be necessary to develop elderly-friendly devices suited for characteristics of the elderly group and to actively deploy professional manpower, in parallel with active support from related organizations, who can provide instant support to the elderly encountering difficulty with the use, so as to stimulate the smart care service in the period ahead. Furthermore, as health information and privacy information generated in connection with individuals are sensitive data, it would be the most important to ensure that strict security is maintained for safe data transmission and that only necessary part of information is shared selectively.

Keywords: Smart care · The elderly · Health care · U-health · Monitoring · Smart home

1 Introduction

As the nation is making transition to an aged society, the health care for the elderly has taken on an added importance and there is a widespread recognition that systematic and efficient management system is required. The key to the technologies required for the elderly health care system would be to obtain and manage various signals from the elderly in an unconstrained and non-intrusive state by using existing stable technologies, rather than cutting-edge technologies, given that the elderly are not easily adaptable to new machinery. The health-related field in the aging society serves as a backbone of future core technologies leading from E-Health to U-Health and smart care. Smart care, which remotely manages health status of the elderly with chronic diseases who need

K.J. Kim et al. (eds.), *IT Convergence and Security 2017*,
Lecture Notes in Electrical Engineering 450,
DOI 10.1007/978-981-10-6454-8_9

constant health management while providing health education and disease management for patients, can be considered to represent a fusion between ubiquitous health and health education [1].

Smart care technology which has made strides is recognized vital in the aging society. Various smart devices used as media for smart care have usefulness in that they enable users to gain access to a variety of information services anytime and anyplace by utilizing information communication. In addition, the absence of constraint on time and space leads to an increase in the ease of accessibility by users, opening up the possibility for improvement of effectiveness and efficiency in the delivery of healthcare services based on convergence between health promotion and management services, as well as disease management. Thus, a sea change is expected to be brought to the environment of health management medical services in the period ahead [2].

Thus, this study was intended to examine trends and outlook of smart care by type of elderly and situation to spur improvement in health management for the elderly.

2 Trend of the Smart Care by Type of Elderly and Situation

(1) Elderly-friendly smart home

The key concept of smart healthcare is to overcome spatial limitations and enable constant health management through application of IT (information technology) to existing residential space of the elderly [3]. The smart home system is an important means to maintain health conditions and quality of life of the elderly and will become indispensable in aging society. Lim and Chung suggested that spatial planning was important for building a smart home [4] and presented technological elements of elderly-friendly smart home based on four categories, i.e., life, health, security, and communication (Fig. 1).

(2) Smart care for management of chronic diseases in the elderly

E-health and U-health have emerged as alternatives to cope with shortage of physicians which arises from population aging and to deliver high quality, sustainable and high-efficient treatments to the elderly. In particular, E-health and U-health will help increase the number of patients receiving sustainable and proper/diagnosis among residents in areas with poor medical facilities and low population density, allowing patients to stay home longer prior to admission to hospitals. Moreover, guidelines may be presented for diagnosis and treatment through evidence-based knowledge system [5]. In addition, constant management of daily lives is far more important than short-term treatment for major chronic diseases, such as diabetes and hypertension [6]. Non-contact and non-pain biometric information measurement has been recently evolving, enabling transmission of data from devices with greater convenience to manage disease status, prevent fall, send out medication intake notification, and transmit notification alerts if a dose is missed, and furthermore, remote medical services will be able to be delivered to monitor health conditions of the elderly patients at home.

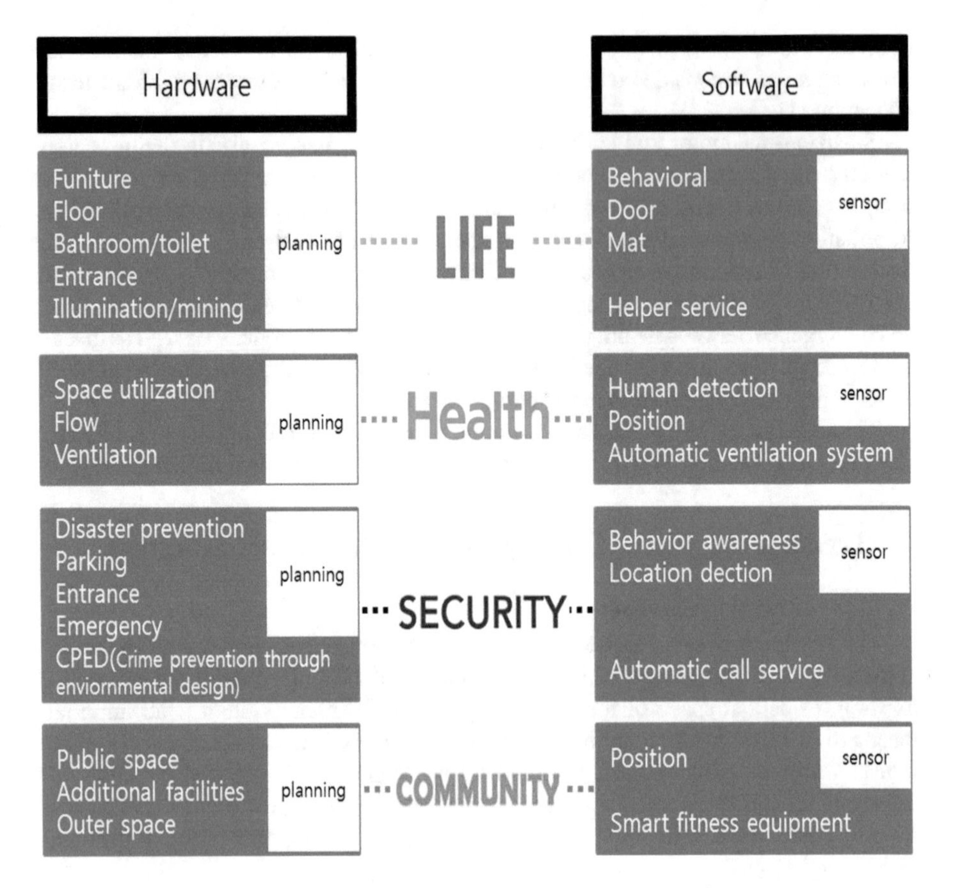

Fig. 1. Elements of an elderly friendly smart home space planning [4]

(3) Smart care for the elderly living alone

The issue of the elderly living alone is likely to arise from recent surge in the number of single-person households and the nation's entry into the aging society. Most elderly living alone are isolated from the outside and have the degraded ability to make decisions in emergency situations and unable to cope with crisis in everyday life. Additionally, problems such as the elderly dying alone, etc., have recently emerged as social issues and therefore care needs to be given to the elderly living alone.

Examples of smart care for elderly living alone in Korea include the 'U-Care System for the Elderly Living Alone' which is operated by Jeollanam-do Provincial Government jointly with Gwangyang-si and 'Emergency Safety Caregiver System for the Elderly Living Alone' which is operated by Busan Metropolitan City. In Japan which has an advanced elderly-related industry, smart care for elderly living alone has been provided, such as 'Living Status System for the Elderly Living Alone' which is operated by the Central Research Institute of Electric Power Industry (CRIEPI) and 'Self-Environment for Life' operated by Tokyo University of Japan.

The smart care for the elderly living alone in Korea has focused on prevention of lonely death of the elderly through health condition tracking, safety monitoring of the elderly, and immediate response to emergencies related to their health conditions. Meanwhile, the Central Research Institute of Electric Power Industry (CRIEPI) introduced the concept of non-intrusive sensing, enabling remote monitoring of the living conditions of the elderly living alone through changes in electric current (Fig. 2).

Fig. 2. CRIEPI's watching system for the elderly living alone [1]

The Self-Environment for Life operated by the Intelligent Cooperative Laboratory at Tokyo University of Japan measures the condition of respiratory organs in the elderly and manages their health condition by using the ceiling dome microphone and pressure sensor bed (Fig. 3). In addition, IoT (Internet of Things) has been recently used for multi-faceted monitoring of the health conditions of the elderly living alone.

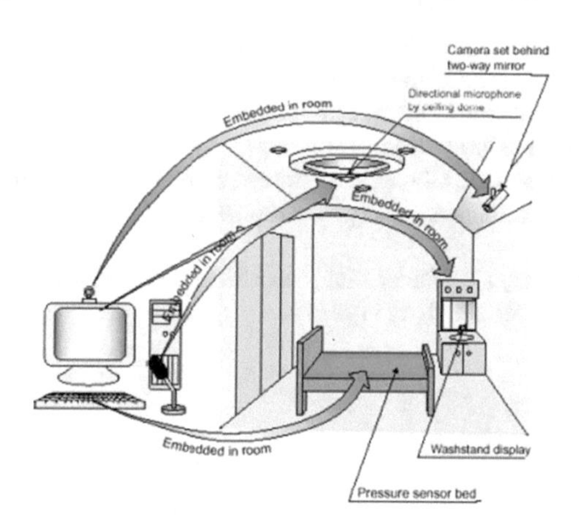

Fig. 3. Diagram of Self Environment for Life [1]

3 Conclusion and Recommendation

Smart care is needed to be implemented actively not only to reduce medical costs and improve the quality of medical services but also to ensure safety and quality of life for the elderly. However, multi-faceted efforts need to be made to help overcome the difficulties arising from degraded cognitive function of the elderly, considering the basic orientation of smart care such as mobile, smart, cloud computing, IoT, etc. The elderly who experienced smart care indicated that it also had disadvantages, such as difficulty with use of measurement devices, difficulty in resolving system disorder or communication disorder, inconvenience in using the system due to presbyopia [7], despite the convenient and interesting features of smart care. For that, it would be necessary to develop elderly-friendly devices suited for characteristics of the elderly group and to actively deploy professional manpower, in parallel with active support from related organizations, who can provide instant support to the elderly encountering difficulty with the use, so as to stimulate the smart care service in the period ahead. Furthermore, the most important issue in smart care market is the security issues. As health information and privacy information generated in connection with individuals are sensitive data, it would be the most important to ensure that strict security is maintained for safe data transmission and that only necessary part of information is shared selectively.

References

1. Kim, K.H.: Managing elderly people living alone with smart care technology. Korean Inst. Electr. Eng. **62**(12), 28–33 (2013)
2. Kim, H.K., Cho, Y.T.: Smart health, leading a change to healthy life: policy issues. Health Welf. Policy Forum **199**, 70–78 (2013)
3. Kim, J.J., et al.: Analysis of planning elements for elderly housing space considering the characteristics of the elderly. Archit. Inst. Korea Conf. Proc. **27**(9), 151–160 (2011)
4. Lim, S.M., Chung, J.H.: A study on the elderly-friendly smart home space planning based on healthcare—focused on user preference. Archit. Inst. Korea Conf. Proc. **34**(2), 113–114 (2014)
5. Ha, E.S.: The implementation of smart care system for dementia patients. J. Korea Acad. Ind. Coop. Soc. **15**(6), 3832–3840 (2014)
6. Cho, S.Y., Lee, S.J., Lee, I.H., Chung, J.H.: Spread of Wellness Care and the Medical System in the Future. Korea Information Society Development Institute, Ministry of Science, ICT and Future Planning (2015)
7. Lee, Y.J., Lee, J.H., Nah, J.Y.: Older adults' experience of smart-home healthcare system. J. Korea Contents Assoc. **5**(5), 414–425 (2015)

Web Technology and Applications

Trip Basket: Web Characterization for Smart Tourism

Wasana Ngaogate(✉)

Ubon Ratchathani University, Ubon Ratchathani, Thailand
wasana.n@ubu.ac.th

Abstract. This paper proposes new features for upgrading existing Web application, called TripBasket, in order to support smart tourism, especially enhancing an individual tourist's experience including elder tourists who regularly travel alone in their young age. Four characterized features are tourist's preference, location-based notification, unique entrepreneurs, and data mining in tourism. Some practical functions are introduced so that the further work can be done practically with cooperation of entrepreneurs.

Keywords: Smart tourism · Location-based service · Innovation · Aging society

1 Motivation

Technologies for tourism are getting more important since emerging of social media and big data on the cloud. Tourists get more convenience along their trips according to the technologies. Stakeholders related to tourism also get benefits, for example, entrepreneurs in local destinations promote their businesses via social network; such as Facebook, Instagram; for free of charge. Both local entrepreneurs and tourists are able to make uses of technologies on their own without being member of some middleman like Booking.com, AirBnB.com, TripAdvisor.com, etc. It might be nothing more features of technologies for tourism unless a term of smart tourism has not come up.

An intelligent advise, for tourists, given by application both Web application and mobile application is new for tourism. This paper therefore reviews some articles in smart tourism in order to find out what intelligent features should be added to existing tourism technologies. Because the author has developed a Web application, called TripBasket, and demonstrated it in one of Thai startup competition in early 2017. The application was designed and developed based on basic needs of tourists under the influence of personal experience of the author who frequently behaves as a solo traveler with no plan.

In order to make the TripBasket more intelligent, literature review in Sect. 2 is studied. New features and characterization is proposed in Sect. 3. Section 4 concludes what has been done and suggests some further works.

2 Research in Smart Tourism

Differences between e-Tourism and smart tourism have been clarified by Ulrike Gretzel, et al., [1] in seven aspects. They defined smart tourism as tourism supported by integrated

K.J. Kim et al. (eds.), *IT Convergence and Security 2017*,
Lecture Notes in Electrical Engineering 450,
DOI 10.1007/978-981-10-6454-8_10

efforts at a destination to aggregate data derived from physical infrastructure, social connections, organizational sources and human in combination. With the use of advanced technologies, smart tourism transforms data into on-site experiences and business value-propositions with experience enrichment. The same as Kim Boes, Dimitrios Buhalis and Alessandro Inversini [2], they claimed that ICT, leadership, innovation and social capital supported by human capital are core components of smartness as it is insufficient on its own to introduce smartness. Destination managers have to acknowledge the multi-facet construct of smartness to create value for all and enhance competitiveness. Therefore, human still need to be important part of smart tourism.

Yunpeng Li, Clark Hu, Chao Huang, Liqiong Duan [3] compared the characteristics of both traditional tourist information services and those incorporated in smart tourism. For the Chinese tourism market, smart tourism represents a new direction implying a significant influence on tourist destinations, enterprises, and also tourists themselves. They concluded that smart tourism is the ubiquitous tour information service received by tourists during a touring process.

Getting information while traveling is widely popular as the Internet service is accessible easily across countries. Research in mobile usability is then one of research topics in tourism. Andréa Cacho, Luiz Mendes-Filho, et al. [4] have tried out a mobile application, called FindNatal, while FIFA World Cup 2014 at Natal city, Brazil. They aimed to assess the feasibility of the Find Natal in order to find out three questions: the main visited places, meeting points among different groups of tourists, and outliers in the tourist flow.

Tracking where tourists have been is also interesting. Noam Shoval and Rein Ahas [5] have reviewed 45 papers in tourism in order to find out the use of tracking technologies. They concluded that no doubt that the future of tracking technologies in tourism because of the rapid advances in tracking technologies and the growing possibilities in implementing them. Weimin Zheng, Xiaoting Huang, Yuan Li [6] presented a heuristic prediction algorithm to predict the next locations of tourists using GPS data collected from 111 tourists at the Summer Palace in Beijing. The experimental results suggest that their proposed algorithm has significantly higher prediction power and accuracy than existing methods.

Data mining is one of outstanding research in smart tourism. Alfredo Cuzzocrea, et al. [7] have proposed a big data analytics framework, called FollowMe, that permits to follow traveling Twitter and Instagram users by tracking their geo-located messages they post during their trips, so that several analysis dimensions (i.e., Time Slot, Origin Airport, Path, and others) can be built over them and exploited to analyze results. Yuan Yifan, Du Junping, Fan Dan, JangMyung Lee [8] applied text mining on Baidu travel notes in order to recognize and discover tourism activity quickly, accurately and conveniently.

Smart destination can be suggested by technology as Cacho A. et al. [9] presented a platform that uses social media as a data source to support the decisions of policymakers in the context of destinations initiatives. The initial results suggest that data collected from Twitter posts can be applicable to the effective management of smart tourism destinations. Whereas Yong Liu, et al. [10] have already analyzed over four hundreds thousands of customer reviews on TripAdvisor on ten thousands of hotels in five Chinese cities. Based on five attributes of hotels which are rooms, location, cleanness, service, and value, Chinese tourists domestically exhibit distinct preferences for room-related hotel attributes compared to foreign tourists. They concluded that having a different language or cultural background affect customers' preference related to hotel attributes. The results suggest that, for tourists from abroad, the service, the room, and value are the key determinants of hotel ratings, followed by cleanliness and location.

3 Trip Basket

3.1 Basic Features

The application was designed and developed based on basic needs of tourist under the influence of personal experience of the author with some basic features for two types of users: tourists and entrepreneurs. The following photo is one of major functions of the TripBasket: writing a diary and gathering several diaries as a new journal.

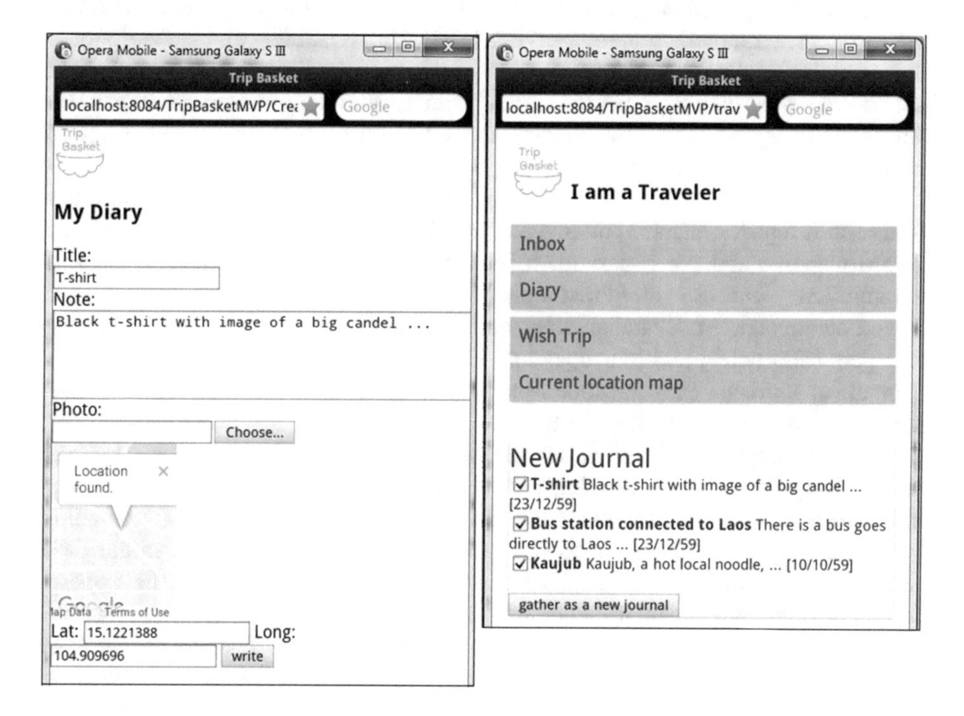

The following are basic features for two types of user.

1. Tourists
   - Communicating with entrepreneurs.
   - Writing diary embedded with location address (Latitude-Longitude) and sharing to social media (Facebook).
   - Gathering diaries into a journal.

- Showing surrounded attractions and businesses on a map.
- Making wish lists of destinations.

2. Entrepreneurs
   - Providing information of their businesses.
   - Communicating with tourists.

3.2 New Features for More Intelligence

According to definition defined by Ulrike Gretzel, et al. [1], smart tourism takes role "during trip" whereas e-Tourism involves with pre and post travel. Therefore, the new features will focus on enhancing tourists' experience during their trips.

The TripBasket implements responsive Web design so that the same content is readable properly on various screen sizes. Terry Cottrell and Brigitte Bell suggested that it is reasonable to expect that users want to move seamlessly among tablet device, laptop, desktop, and other devices so that their experience will be consistent.

Weimin Zheng, Xiaoting Huang, Yuan Li [6] proposed that they can expand their work from location prediction for individual tourists to predicting tourist distributions. However, the TripBasket is not designed for destination prediction as it allows individuals to make their own wish list for particular area before they start travel so that they can follow their plan easily. Destinations in the wish list can be introduced during the trip as location-based service.

As entrepreneurs, they have important role in promoting destinations. Suosheng Wang and Joseph S. Chen [11] discovered that place identity affects resident's attitudes toward negative and positive tourism impacts. A sense of uniqueness or sense of continuity may impact upon tourist's place-based self-esteem and self-efficacy. Paraskevi Fountoulaki, M. Claudia Leue, and Timothy Jung [12] interviewed twenty managers from hotels and tour operators in Crete city at Greece, about distribution channels of tourism. They identified that social media and a mobile location-based service is important for market distribution.

Not only uniqueness, but branding is also important for destination. Sheng-Hshiung Tsaur, Chang-Hua Yen and Yu-Ting Yan [13] paid attention on branding of destination and focus on the identity. They identified five dimensions of destination, for branding, which are image, quality, personality, awareness, and culture. Tourists can use those dimensions to identify differences of destination characteristics. Therefore, the proposed dimensions can differentiate destination from competitors.

According to location-based feature of the TripBasket, new features should make more benefits related to current location of a particular tourist. New features are suggested as following.

- Enhancing tourists experience during their trips by giving location-based suggestion.
- Supporting entrepreneurs to make uniqueness of their businesses by distinguishing destination's identity along with promoting branding based on five dimensions of destination: image, quality, personality, awareness, and culture.

- Collecting data of tourists in three phases: before, during, and after the trip. It will be used for data mining in order to provide more accurate suggestion for particular tourist.

3.3 Characterization of New Features

In order to clarify proposed features of the TripBasket, this subsection describes characters of the new features.

1. **Tourist's preferences.** Not only general information; such as gender, age, etc.; but also keywords of a particular tourist are stored as his/her own preference. Accommodated suggestions for individuals could be easily proposed. They can also have their own plan for the whole trip.
2. **Location-based notification.** Information is fed from articles, diaries, or social network related to current location of a particular tourist. Some conditions are predefined, for example, distance, high ranking, or close to personal preferences. Therefore, the tailored information is pushed to a particular tourist at the right location during the trip. An individual tourist can therefore get on-site experience.
3. **Unique entrepreneurs.** In order to enhance destination identity along with branding, the system asks them to fill in the following information: identity of their businesses related to destination, images, quality compared to standard, unique personality, focus awareness, and culture of both the business and destination. Such identity is one of filter factors for accommodating individuals.
4. **Data mining.** The TripBasket utterly aims to find out similarity of tourists by mining collected data of a particular tourist. Information provided by the system could be more accurate be feeding the right information at the right time at particular location.

3.4 Example Functions of the Smarter TripBasket

Merging with original basic functions of the TripBasket, some smarter functions are proposed as below in order to support tourists before, during, and after the trip.

1. **Tourists**
   - Setting personal profile.
     - General information: such as gender, age, etc.
     - Keywords of interest: such as culture, slow life, mountain, etc.
     - Marking articles or diaries of other users in the system as favorites.
   - Looking surround attractions and businesses on a customized map particularly generated for an individual tourist. The map can be generated as either real-time location-based or preferred location-based.
   - Communicating with entrepreneurs who are suggested by the system or pre-selected by the tourist.
   - Writing diary embedded with location address (Lat-Long) and sharing to social media.
   - Gathering diaries into a journal and publishing with some keywords for mining by others and the system.

- Making wish lists from destinations proposed by using data mining technique, for example, destinations of similar tourists classified by the system.
- Supporting elder tourists as many of them usually travel on their own since they were young. They are more technologically [14] than they used to be and have confidence and independence in traveling.

2. **Entrepreneurs**
   - Providing information of their businesses with unique identity. The identities of all businesses in the system can be classified and filtered for proposing to candidate tourists who have similar characters to identities. Ultimately, the system could provide suitable destinations to a particular tourist.
   - Communicating with tourists before, during, and after the trip, for example, proposing suitable promotions for individuals, sending instant supporting messages while the trip, and summarizing feedback for further service improvement.

4 Conclusion and Further Works

This paper has introduced a Web application for tourism, called TripBasket, demonstrated in Thai startup competition in early 2017. It was recommended to be improved in order to be distinct from existing application. Research papers in smart tourism were studied in order to uncover what features could make the TripBasket smarter especially enhancing an individual tourist's experience. New features are then proposed and characterized in order to clarify how the system will be developed. By the way, existing functions still remain.

Aging tourism is quite new as elders are encourage to have better quality of life. [15]. Kam Hung and Jiaying Lu [16] suggested that understanding aging in the tourism and hospitality contexts can be achieved in two directions: the elderly as a traveler and the elderly as a member of the tourism community.

The next job is iteratively designing user interface and user experience (UI&UX) with cooperation of entrepreneurs until satisfaction is met. Then software development process, such as database design, coding, and testing, will be done. The TripBasket could be tried out at one local destination at the beginning and expand to other area later.

Acknowledgement. This research is supported by Ubon Ratchathani University in the project of researcher enhancement 2017.

References

1. Gretzel, U., et al.: Smart tourism: foundations and developments. Electron. Markets **25**, 179–188 (2015)
2. Boes, K., Buhalis, D., Inversini, A.: Smart tourism destinations: ecosystems for tourism destination competitiveness. Int. J. Tour. Cities **2**(2), 108–124 (2016)
3. Li, Y., Hu, C., Huang, C., Duan, L.: The concept of smart tourism in the context of tourism information services. Tour. Manage. (2016, in press). Available online

4. Cacho, A., Mendes-Filho, L., et al.: Mobile tourist guide supporting a smart city initiative: a Brazilian case study. Int. J. Tour. Cities **2**(2), 164–183 (2016)
5. Shoval, N., Ahas, R.: The use of tracking technologies in tourism research: the first decade. Tour. Geogr. **18**(5), 587–606 (2016)
6. Zheng, W., Huang, X., Li, Y.: Understanding the tourist mobility using GPS: where is the next place? Tour. Manag. **59**, 267–280 (2017)
7. Cuzzocrea, A., Psaila, G., Toccu, M.: An innovative framework for effectively and efficiently supporting big data analytics over geo-located mobile social media. In: Proceedings of the 20th International Database Engineering and Applications Symposium, IDEAS 2016, pp. 62–69 (2016)
8. Yifan, Y., Junping, D., Dan, F., Lee, J.: Design and implementation of tourism activity recognition and discovery system. In: 12th World Congress on Intelligent Control and Automation (WCICA), 12–15 June 2016
9. Cacho, A., et al.: Social smart destination: a platform to analyze user generated content in smart tourism destinations. In: Rocha, Á., Correia, A., Adeli, H., Reis, L., Teixeira, M.M. (eds.) New Advances in Information Systems and Technologies. Advances in Intelligent Systems and Computing, vol. 444. Springer, Heidelberg (2016)
10. Liu, Y., et al.: Big data for big insights: investigating language-specific drivers of hotel satisfaction with 412,784 user-generated reviews. Tour. Manag. **59**, 554–563 (2017)
11. Wang, S., Chen, J.S.: The influence of place identity on perceived tourism impacts. Ann. Tour. Res. **52**, 16–28 (2015)
12. Fountoulaki, P., Leue, M.C., Jung, T.: Distribution Channels for Travel and Tourism: The Case of Crete Information and Communication Technologies in Tourism (2015)
13. Tsaur, S.-H., Yen, C.-H., Yan, Y.-T.: Destination brand identity: scale development and validation. Asia Pac. J. Tour. Res. **21** (2016)
14. Klimova, B.: Senior tourism and information and communication technologies. In: Advanced Multimedia and Ubiquitous Engineering. Lecture Notes in Electrical Engineering (2017)
15. Kim, H., Woo, E., Uysal, M.: Tourism experience and quality of life among elderly tourists. Tour. Manag. **46**, 465–476 (2015)
16. Hung, K., Jiaying, L.: Int. J. Hosp. Manag. **53**, 133–144 (2016)

Trust Blog Ranking Using Multi-Criteria Decision Analysis AHP and TOPSIS

Nurul Akhmal binti Mohd Zulkefli, Baharum bin Baharudin(✉), and Abas bin Md Said

Computer and Information Science, Universiti Teknologi PETRONAS, Teronoh, Malaysia
nurul@fskik.upsi.edu.my, baharbh@gmail.com, abass@utp.edu.my

Abstract. Following the lack of trust for information description and trust support for the query processing, traditional blog systems are unable to satisfy users in terms of the performance of information organisation and retrieval. As well as the fact that the information provided by the website is undefined trust and a user keeps on searching the information from the internet, it is important for a user to have some belief or trust in the information they had read. The multi criteria decision making technique (MCDM) through the integration of the analytic hierarchy process (AHP) and technique for order preference by similarity to ideal solution (TOPSIS) method is introduced to provide the trusted score for ranking the blog based on the blog trustworthiness level. Our MCDM methods are developed to consider the important keys of trusted information, which entail great agreement from experts based on the proposed criteria which are the follower, viewer and post. We present experimental results that can beneficially be used by the user whenever they are looking for the blog information and they can manage to rank the result based on the trusted score value.

Keywords: Trust · Blog · Multi-Criteria Decision Making · AHP · TOPSIS

1 Introduction

Blogs are one of the most popular Web 2.0 platforms and can be classified as personal (or organizationally unaffiliated) or organization sponsored (e.g., a museum-sponsored blog). Personal blogs are "diary-style web sites" on which bloggers post their opinions, reviews, ideas, personal stories, and emotion. Nowadays blogs occupy an increasingly important place in the information worldwide. Whilst sharing the knowledge and experience, they also become automatically involved in a communication that exists between blogger and visitors via comments or feedback. In the blog circle life, the information is kept updated, something which is known as real-time information. However, another interesting thing about the blog is that we cannot be sure if the blog can be trusted or not. It is good to understand that blogs contain user experience information and suitable as alternative information source for people but if people know if the blog information is trusted or not well trusted, it will be more efficient and the result will be more satisfied.

K.J. Kim et al. (eds.), *IT Convergence and Security 2017*,
Lecture Notes in Electrical Engineering 450,
DOI 10.1007/978-981-10-6454-8_11

The majority of the current work on the blog recommendation are focusing on the tagging system, keywords, user profile and categories. The information is getting larger every day and it is difficult to maintain some methods such as the computation for tagging similarity, the similarity of keywords, ontology database and user-profile maintenance. Besides, there has been less focus on trust information in blog recommendation. Trust is everything in the online world and according to a recent study by [1] in Econsultancy, 61% of customers read and trust online reviews when making a purchase. Meanwhile, according to [2] in Evercoach, 81% of U.S online consumers trust the information on blogs. Meanwhile, [1] mentioned that nowadays travel blogger has their own personality and our readers trust our recommendations because they relate to us. Besides, [4] stated that a travel blogger can reach people in a way that is more effective, engaging and often less costly than traditional media. The research aim of this study is to look further into some novel, effective and efficient approaches to improve the quality of blog recommendation by means of implementing the trust and exploring a new alternative criteria for ranking other than common rating and sorting information. In order to improve the recommendation result by implementing the trust score, we are interested to introduce one of the rich information contained on the internet namely blog. To provide a broader scope of evaluation, we proposed new measurement for trustworthiness in blog. The trustworthiness is defined by having a trust score for each blog using three proposed criteria from blog which is the total number of follower, viewer, and post. All these criteria are discussed in next section.

2 Literature Review

Personal blogging in Malaysia had come to peak where the relatively stabilized conventions and expectations of the genre had emerged to sustain persistent interpersonal and social dynamics. Malaysian Communications and Multimedia Commission (MSMC) had mentioned about the Internet Users Survey 2016 [3] that interviewed a total of 2,787 respondents through Computer Assisted Telephone Interview (CATI) system. MSMC asserted that the internet remained to be an important source of information for 90.1% users, while 80.2% said they were totally addicted on social media. As trust is a social phenomenon, the model of trust for an artificial world like the Internet should be based on how trust works between people in society [4]. The rich and ever-growing selection of literature related to trust management systems for Internet transactions, as well as the implementations of trust and reputation systems in successful commercial applications such as eBay[1] and Amazon[2] give a strong indication that this is an important technology. One of the researchers is [5] recommend a blog recommendation mechanism that combines trust model, social relation and semantic analysis and then illustrates how it can be applied to a prestigious online blogging system with the emphasis on Taiwan. From their experimental study, they found a number of implications from the Weblog network and several important theories in the social networking domain that is empirically justified and their proposed recommendation mechanism is

---

[1] https://www.ebay.com/.

[2] https://www.amazon.com/.

claimed to be feasible and promising. ITrustU is an online blog article recommender developed by [6]. This recommender is used to ease the information overload in the blogosphere with a trust enhance collaborative filtering approach that integrates multi-faceted trust following the article type and user similarity. In our research, we proposed the blog trust ranking by using the trust weight from three criteria of blog. One of the main reasons these criteria are chosen is because it shows the numbers that is actually showing the truth or real of the blog. [7] Mentioned that flaunting comment numbers, share count, subscriber count, etc. are all acts as trust triggers for your readers. Such numeric figures can act as social proofs and can give a significant impact on ranking the blog's trustworthiness. The criteria are described details in next section.

The Multi-Criteria Decision Making (MCDM) was developed since 1970s as one prominent model to measure and evaluate most of the chosen criteria. The MCDM models are deemed appropriate for evaluating and making decision for the best alternative options in order to select the perfect criteria [14]. By adopting a decision theory, multi-criteria systems can prepare or make available a lot of rich tools for system designers to build more interesting systems as well [15]. It is also worth nothing that allowing online visitors to provide fine grained multi-criteria rating feedback is common in the travel and tourism industry. Recommender systems and decision support systems have been trying to help the user and decision maker throughout the recommendation and decision making process, respectively [16]. Thus, recommender systems can be considered as decision support systems where a target user in a recommender system can respond to a decision maker in a decision support system. In the decision making theory, MCDM helps the decision maker in the decision making process when multiple criteria (i.e. variables, attributes etc.) conflict and stands in competition with each other. Most commonly used decision-aiding methods, such as outranking methods and the value-focused models, depend strictly on the multi-criteria aggregation procedures. In this paper, the integration of AHP and TOPSIS is used to get the trustworthy blog. The Analytic Hierarchy Process (AHP) algorithm can be used to make pairwise comparisons among criteria. AHP is one of the most popular MCDM methods [16, 17]. This method is used to solve a complex decision making problem having several attributes by modeling unstructured problem under study into hierarchical forms of elements. TOPSIS (Technique for Order Preference by Similarity to Ideal Solution), developed by [18] is functionally associated with problems of discrete alternatives. It is one of the most practical techniques for solving real-world problems. TOPSIS attempts to indicate the best alternative that simultaneously has the shortest distance from the positive ideal solution and the farthest distance from the negative ideal solution [19].

3 Methodology

In this section, there is an evaluation and selection of the blog that involves procedures and steps to choose the blog as follows:

3.1 Preliminary Investigation of Availability Blogs

In this paper, the investigation of availability blogs only apply to travel scope. As we know, each blog has its own categories such as politic, gossip, education however travel category is one of the most top growing category. Overall, we selected 411 available blogs based on access date at June 2015 to December 2015. These blogs are collected using specific keyword such as "travel to Kuala Lumpur" and we choose the result provided by a Google from the 1st page's result to 10th page's result as it was mentioned by the Google that the result after 9th pages are not highly recommended.

3.2 Evaluation – Establish the Criteria for Evaluate Trustworthiness Blog

In this stage, the criteria for trust blog recommendation making which is follower, viewer and post are identified. The details of each criteria are described as below:

1. *Follower:* From the Cambridge Dictionary definition means that someone who has a great interest in something and someone who supports, admires, or believes in a particular person, group or idea [8]. Meanwhile, Blogspot defines *follower* as someone or readers who are a fan of what the blogger has posted. Among the desirable *follower* outcomes posited to arise from authentic leadership and followership are heightened levels of trust [9, 10]. Moreover, from the book "The Trustworthy Leader" by [11], the author identifies six elements that reflect a leader's trustworthiness: honor, inclusion, engaging followers, sharing information, developing others, and moving through uncertainty to pursue opportunities.
2. *Viewer:* One of the trust trigger by [12] is Subscriber/Follower/Reader Numbers where he mentioned that similar to blog comments, the number of people who subscribe, follow or read your site can help you build trust instantly. The author also agree that the blogger should show whatever number that proves blogger's credible because when the numbers are small and insignificant, it can have the reverse effect and erode it instead of building credibility. According to [13] some high pageview bloggers could have a great influence on a consumer's shopping decision. So, in our research we assume that with a large number of *viewers*, the blog itself can have the one of criteria trustworthiness.
3. *Post:* or entry is the main content of the blog and it depends on how blogger presenting the content inside his/her blog. In our research, we used the total number of *post* provide by the blogger. We choose to have the number of *post* as because it shows how active is the blogger in post the content. We believe that if blog have many *post* than the blog have more trust score and more trustworthy content compare to few number of *post* in blog. One of the *post* shared by expert blogger in the post title "Build Trust" is *'Provide great content consistently'* where the author mentioned that 'if there are no regular content on your blog, it looks stale of your readers and when you publish regular content on your site, it acts as a trust factor', by [7].

3.3 Selecting Trustworthy Blog

This stage includes the collected of criteria weight in the metric table using Analytic Hierarchy Process (AHP) and based on each criteria weight assigned by expert, the available blogs are ranked in descending order of the score. Technique for the Order of Prioritization by Similarity to the Ideal Solution (TOPSIS) is used to rank the best trusted blog. Using aggregation in TOPSIS, the ranking for trustworthy blogs are shown however the result may be vary as it can rely on individual judgements.

3.4 Determining the Criteria Weight by AHP and TOPSIS

The AHP measurement matrix is processed to obtain the weights based on the evaluators' preference. The stepwise procedure [17] is used for implementing TOPSIS. The first step start with the construction of the normalized decision matrix and step 2 by constructing the weighted normalized decision matrix. In step 3, the ideal and negative ideal solutions is determined using, meanwhile in step 4, separation measurement is calculate based on Euclidean distance. For step 5, closeness to the ideal solution is defined and finally, in the last step, the ranking of the alternative is determined according to the closeness to the ideal solution when the set of alternative is now ranked according to the descending order of and the highest value indicates the best performance.

4 Result and Discussion

Each result and analysis are discussed in two categories which is category by AHP and category by TOPSIS. In AHP, all evaluators result are compiled to get the final weights for each criteria and in TOPSIS, these weights are used to rank the trustworthiness of blog. Details are explained in the sections below. Table 1 shows the example of 20 collected data blog from 411 blogs dataset.

Table 1. The assessment of a sample of blog data based on *followers*, *posts* and *viewers*

Original blog data samples							
Blog. no	Follower	Viewer	Post	Blog. no	Follower	Viewer	Post
1	1	1020171	1445	21	0	35423	74
2	1864	2584655	530	22	3090	0	828
3	0	0	0	23	3116	0	811
4	0	0	0	24	383	1048742	231
5	0	0	19	25	57	87632	452
6	0	5	0	26	0	0	146
7	82	0	99	27	121	6208	77
8	0	0	50	28	6920	0	1940
9	129	254611	217	29	4	0	50
10	0	0	798	30	61220	4494888	2090
11	6716	423682	191	31	525	299169	1321
12	0	0	2	32	15455	4306424	0
13	144	0	187	33	41	841706	95
14	117	0	2661	34	118	123348	158
15	2190	419935	1299	35	159	0	0
16	1618	1235061	339	36	64	3255	34
17	6	151986	112	37	8	0	35
18	0	0	307	38	1272	4285518	1970
19	79	692353	959	39	0	0	44
20	16789	5269974	195	40	49	79634	766

4.1 Weights Criteria by AHP

The results of the analysis (Table 2) were then used to build a decision matrix. Essentially, AHP is a technique or a process that obtains ratio scales from paired comparisons, allowing small inconsistencies in judgements due to inherent human errors. For such technique, several experienced evaluators are needed to evaluate several criteria required for the analysis. For this study, eight evaluators were recruited to perform a comparative analysis on three criteria. Ideally, such evaluators should be industrial practitioners and academic experts, with relevant knowledge and experience. The final AHP results were then evaluated against the importance of each criterion, as shown in Table 2. The results of the AHP analysis showed that the subjective judgements made by the evaluators had high consistencies (CR) where consistency for each evaluator, CR are 0.02, 0.96, 0.87, 0.00, 0.36, 0.90, 0.65, and 0.19, respectively. Clearly, such CR values were smaller or equal to 1, which further strengthened the reliability of the trust score results

Table 2. The results of AHP pairwise comparison in selecting trusted blogs by eight evaluators

Evaluator	Follower (A) vs. viewer (B)	Follower (A) vs. post (B)	Viewer (A) vs. post (B)
1	B is very strongly important than A	B is strongly important than A	A is equally important as B
2	A is very strongly important than B	A is slightly important than B	A is strongly important than B
3	A is strongly important than B	B is slightly important than A	B is very strongly important than A
4	A is equally important as B	A is equally important as B	A is equally important as B
5	A is very strongly important than B	A is strongly important than B	B is slightly important than A
6	B is extremely important than A	B is strongly important than A	A is strongly important than B
7	B is very strongly important than A	B is strongly important than A	B is strongly important than A
8	A is strongly important than B	B is strongly important than A	B is slightly important than A

4.2 TOPSIS Ranking Results and Discussion

Table 3 shows the mean scores of blogs based on the internal and external aggregations in the overall TOPSIS analysis. Clearly, such analysis using both external and internal aggregations had resulted in B_{144} being chosen as the first most trusted blog, which was followed by B_{133} and B_{346} as the second and third most trusted blogs, respectively. Table 4 summarizes the ranking of blogs by the eight evaluators in this study. Evidently, the proposed criteria, namely *follower*, *viewer* and *post*, as measures or weights of trust for blogs are deemed appropriate, given the consistent ranking of blogs by all evaluators. For example, blogs B_{144}, B_{133}, and B_{346} were consistently ranked by the eight evaluators as the top three most trusted blogs (in the descending order). Furthermore, even though the rankings of the remaining blogs by the evaluators are not the same, the blogs nevertheless remain in the list of the top 10 trusted blogs. The proposed new criteria-based method to determine the trust scores of travel blogs). Essentially, this is a hybrid method, integrating AHP and TOPSIS techniques, to analyze and compute the trust scores of blogs using three criteria of trust, namely *follower*, *viewer* and *post*. Such a focus on these three criteria departs from the usual emphasis on other measures commonly used by other methods, such as similarity tagging, user-based profiles and item-based trust weights.

Table 3. Group decision making of TOPSIS with internal and external aggregation for Top 10 trusted blog

External aggregation					Ranking	Internal aggregation			
Blog. no	Follower	Viewer	Post	Score		Blog. no	S^+	S^-	Score
144	1101774	0	215	0.705478	1	144	1.806175	4.429194	0.710334
133	559236	12521352	0	0.456969	2	133	2.714219	2.308658	0.459629
346	4582	37765998	2726	0.235546	3	346	4.62818	1.41139	0.23369
231	61759	3124526	22524206	0.201502	4	231	4.421389	1.113403	0.201164
342	147620	14160420	375	0.164133	5	342	4.183928	0.815552	0.163127
197	8919	11966542	2067	0.088763	6	197	4.726036	0.448919	0.086748
292	103761	0	0	0.08395	7	292	4.502996	0.417125	0.084779
343	1926	9601496	2315	0.071674	8	343	4.773892	0.358892	0.069921
156	4518	8960094	0	0.06735	9	156	4.770251	0.335423	0.065696
30	61220	4494888	2090	0.062366	10	30	4.604839	0.304997	0.06212

Table 4. The top 10 trusted blog ranking by the eight evaluators

Trusted blog ranking	Blog no.							
	Evaluator 1	Evaluator 2	Evaluator 3	Evaluator 4	Evaluator 5	Evaluator 6	Evaluator 7	Evaluator 8
1	B_{144}	B_{144}	B_{144}	B_{231}	B_{144}	B_{144}	B_{144}	B_{144}
2	B_{133}	B_{133}	B_{133}	B_{144}	B_{133}	B_{133}	B_{133}	B_{133}
3	B_{231}	B_{346}	B_{346}	B_{341}	B_{346}	B_{346}	B_{346}	B_{346}
4	B_{342}	B_{231}	B_{342}	B_{133}	B_{342}	B_{342}	B_{342}	B_{342}
5	B_{346}	B_{342}	B_{231}	B_{342}	B_{231}	B_{231}	B_{231}	B_{231}
6	B_{292}	B_{197}	B_{197}	B_{197}	B_{292}	B_{292}	B_{292}	B_{292}
7	B_{245}	B_{292}	B_{343}	B_{343}	B_{197}	B_{197}	B_{197}	B_{197}
8	B_{30}	B_{343}	B_{156}	B_{154}	B_{245}	B_{245}	B_{343}	B_{245}
9	B_{181}	B_{156}	B_{302}	B_{202}	B_{30}	B_{30}	B_{156}	B_{343}
10	B_{229}	B_{302}	B_{292}	B_{20}	B_{343}	B_{343}	B_{32}	B_{30}

More specifically, this criteria-based method capitalizes on existing contents or elements of blogs, thus making it more efficient and effective. As such, this method will be able to calculate and generate trust weights to which the ranking of blogs will be based on. Once blogs are ranked, users can select reliable and current travel information from such blogs.

5 Conclusion and Future Work

In this study, three criteria of trust, namely follower, view, and post, were used in the ranking analysis of 411 selected travel blogs by the recommendations system. The evaluation of the impact of the proposed criteria and hybrid method on blog recommendations is explained. The experimental results showed that the proposed multi-criteria hybrid method based on MCDM technique was both effective and efficient in ranking blogs based on the computed trust weights as compared to the existing conventional methods. Hence, such ranking can help guide online users to choose information that not only current but also reliable. Specifically, a hybrid of the Multi-Criteria Decision Making (MCDM) technique was utilized to integrate the AHP and TOPSIS methods in computing the criteria's weights and in ranking the blogs. Interestingly, the results of the analyses were found to be highly consistent and reliable. Such findings strongly reinforce the validity and reliability of the three criteria (i.e., follower, view, and post) as measures or weights of trust of blogs, especially travel blogs.

References

1. Charlton, G.: Ecommerce consume reviews: why you need them and how to use them. https://econsultancy.com/blog/9366-ecommerce-consumer-reviews-why-you-need-them-and-how-to-use-them/
2. Nawalkha, A.: Six proven ways to constantly create new clients. http://blog.evercoach.com/6-proven-ways-constantly-create-new-clients/#sthash.Qc4nlFYD.dpbs
3. Malaysian Communications and Multimedia Commission: Internet Users Survey 2016 (2016)
4. Abdul-rahman, A., Hailes, S.: Supporting trust in virtual communities. In: Proceedings of the 33rd Hawaii International Conference on System Sciences, Hawai, HI (2000)
5. Li, Y.-M., Chen, C.-W.: A synthetical approach for blog recommendation: combining trust, social relation, and semantic analysis. Expert Syst. Appl. **36**, 6536–6547 (2009)
6. Peng, T.-C., Chou, S.T.: iTrustU: a blog recommender system based on multi-faceted trust and collaborative filtering. In: Proceedings of the 2009 ACM Symposium on Applied Computing, pp. 1278–1285. ACM, New York (2009)
7. Hallur, A.: How to build trust with your readers as a blogger? https://www.gobloggingtips.com/build-trust/
8. Cambridge: Follower. http://dictionary.cambridge.org/dictionary/english/follower
9. Dirks, K.T., Ferrin, D.L.: Trust in leadership: meta-analytic findings and implications for research and practice. J. Appl. Psychol. **87**, 611–628 (2002)
10. Jones, G.R., George, J.M.: The experience and evolution of trust: implications for cooperation and teamwork. Acad. Manag. Rev. **23**, 531–546 (1998)
11. Lyman, A.: The Trustworthy Leader: Leveraging the Power of Trust to Transform Your Organization. Jossey-Bass, Hoboken (2012)
12. Halpern, D.: How to build your blog's credibility in just a few seconds with trust triggers. http://diythemes.com/thesis/trust-triggers/
13. Wu, W., Lee, Y.: The effect of blog trustworthiness, product attitude, and blog involvement on purchase intention. Int. J. Manag. Inf. Syst. **16**, 265–276 (2012)

14. Aziz, N.F., Sorooshian, S., Mahmud, F.: MCDM-AHP method in decision makings. ARPN J. Eng. Appl. Sci. **11**, 7217–7220 (2016)
15. Lakiotaki, K., Matsatsinis, N.F., Tsoukias, A.: Multicriteria user modeling in recommender systems. IEEE Intell. Syst. **26**(2), 64–76 (2011)
16. Shambour, Q., Lu, J.: Integrating multi-criteria collaborative filtering and trust filtering for personalized recommender systems. In: IEEE SSCI 2011—Symposium Series on Computational Intelligence—MCDM 2011 IEEE Symposium on Computational Intelligence in Multicriteria Decision-Making, pp. 44–51 (2011)
17. Hwang, C.-L., Yoon, K.: Multiple Attribute Decision Making. Springer, Heidelberg (1981)

SCADA/HMI Systems for Learning Processes of Advanced Control Algorithms

Christian P. Carvajal(✉), Leonardo A. Solís(✉), J. Andrés Tapia(✉), and Víctor H. Andaluz(✉)

Universidad de las Fuerzas Armadas ESPE, Sangolquí, Ecuador
chriss2592@hotmail.com, leonardosolisc@gmail.com,
andres23138@gmail.com, vhandaluzl@espe.edu.ec

Abstract. This work focuses on the implementation of interfaces for human machine interaction (HMI) for the control of a three-phase motor within an industrial process, using different softwares with which communication can be established to control, monitor and manipulate the variables Which intervene in an industrial process. For this, a didactic module is built using a programmable logic controller (PLC), touch screen and frequency variator for the control and monitoring of a three-phase motor.

Keywords: HMI · Interface · Software · Control

1 Introduction

Throughout the times the industry has evolved thus obtaining greater advantages to automate processes [1, 2], with the automatic revolution the electronic controllers have managed to break several limitations in the petroleum industry, food industry, power generators, among others with which security, flexibility and economy in the operation were achieved, where control methods have been used which Have evolved since [3]: *(i) ON-OFF control* which is the simplest way to control a dynamic variable where the actuator only has two states [4]; *(ii) PID controllers* that actually in the industry have chosen a high percentage for their simple structure, precision, reliability and robustness to a certain extent, with high stability in first order system and in processes where the response times are not so relevant [5]; y *(iii) advanced controllers* [6], are characterized by manipulating multiple variables at the same time, focused on the future prediction of the variables to be controlled, thus presenting systems with greater robustness [7, 8], taking advantage of techniques that integrate mechanics, electronics, and computer science, replacing classic systems. In order to have these characteristics within the industry, the main building blocks as: *(i)* programmable logic controllers (PLC) *(ii)* the field of SCADA devices and among others, need to be flexible and consistent according to their functionality [9].

SCADA systems are increasingly using new computer technologies in the development HMI [7, 9, 10], using sophisticated software with the industrial capabilities necessary to satisfy the process and operator requirements *e.g.* in [9] use web technologies to monitor the level of a tank where not practically the programmable logic

K.J. Kim et al. (eds.), *IT Convergence and Security 2017*,
Lecture Notes in Electrical Engineering 450,
DOI 10.1007/978-981-10-6454-8_12

controller intervenes directly with the web page, but can read and write the variables in the SQL database, this type of technology is implemented by the advantages it provides, such as *(i)* adaptability *(ii)* support for smart deviced *(iii)* web technologies in continuous evolution, among other; in [11] Implement an HMI using LabView software for the monitoring of an autonomous vehicle guided by GPS, where the distance, speed, direction and other variables of interest are monitored in order to identify the proper functioning of the vehicle; in [12] employing an HMI, development in an application known as WinCC (Windows Control Center) using a SQL database, the advantage of this interface is that it allows the visualization and handling of processes, manufacturing lines, machines and installations. The volume of functions of this modern system includes the issuance of event notices in a form suitable for industrial application [13]. As described the HMI are common tools and necessary to be able to interact with the different industrial systems or processes in which a monitoring is necessary [2, 9, 14], In the present work a complete system is built which integrates stages of instrumentation, control and visualization of industrial processes in differents HMI's created by making use of softwares and technologies that continue to advance and revolutionize the industrial field. A system has been created in which the operation of the control can be deployed from one of the implemented HMI.

This work consists of 5 Sections including the Introduction. Section 2 shows the construction and programming of the PLC for the module. Section 3 describes the HMIs implemented in this module for the monitoring and control of the process. Section 4 shows the experimental results of the control and visualization of the process data in each of the HMIs. Finally, the conclusions are presented in Sect. 5.

2 Construction and Programming of the Process

2.1 Construction

In order to comply with the functional requirements of the module, it is considered that the equipment to be used in the control module is industrial, in order to have a system related to an industrial process, Fig. 1 presented the block diagram of the implemented system.

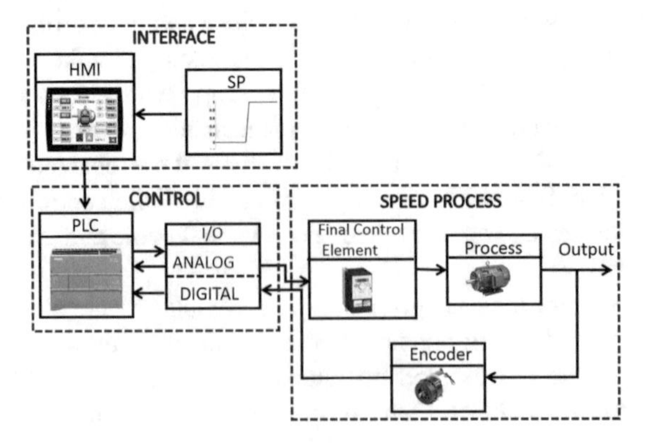

Fig. 1. Block diagram of the speed processing module.

Was studed several modules developed that control the different industrial processes such as; *(i)* temperature, *(ii)* pressure, *(iii)* level, *(iv)* flow rate, *(v)* velocity, among others. The most common ones for research are those that do not require additional equipment for their operation e.g. Pressure, temperature and level using a proportional pneumatic valve, a compressor is necessary, for this reason the speed process is chosen. The piping and instrumentation diagram (PI&D) of the process is represented in the Fig. 2.

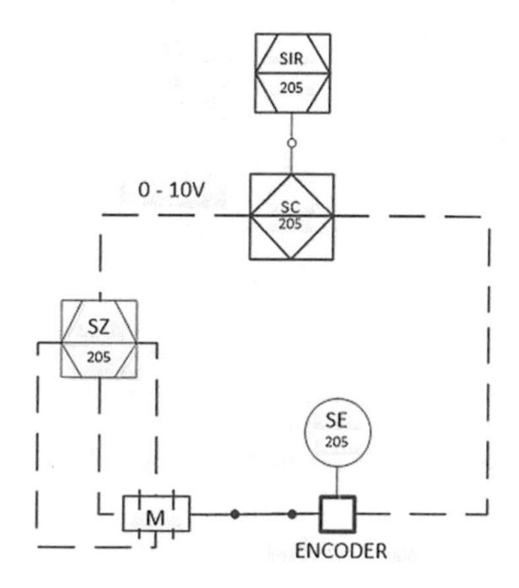

Fig. 2. PI&D of the speed system.

The speed control process has an encoder, which measures pulses and sends a digital signal to the PLC in which calculations are performed to determine the engine RPM. Internally in the *(i)* PLC is developed a PID control algorithm, which has as input the value of the SP in RPS, *(ii)* the actuator is a variable frequency drive that supports the proportional regulation the RPS of the motor.

2.2 PLC Programming

The programming of the PLC employed was done in Ladder language, practically consists of the constant reading of the encoder thus obtaining the process value; Afterwards the execution of the PID loop is generated generating a control variable sent to the actuator, this procedure is executed in a cyclic and continuous way for the process, Fig. 3 shows a flow diagram of the programming of the programmable logic controller.

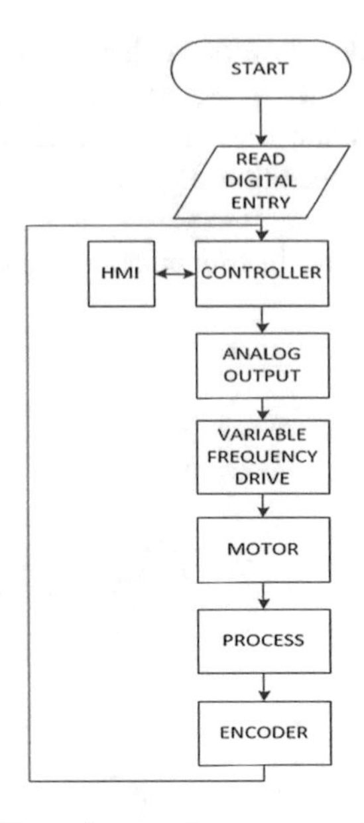

Fig. 3. Flow diagram the program of the PLC.

3 Design of HMI's

A computer-aided HMI is part of the computer program that communicates with the user. The term user interface is defined as all parts of an interactive system (software or hardware) that provide the information and control necessary for the user to perform a task with the interactive system.

3.1 Touch Panel Interface

For the implementation of the HMI in a touch panel, the software TIA PORTAL is used that allows the programming of the screen of the Siemens brand, this software has the necessary requirements to develop an interface in which the values of the process and of Similar way to write the value of get of the RPS of the engine. The HMI used on the screen is connected to the programmable logic controller wirelessly via ethernet either directly with the plc or to a switch as shown in Fig. 4; For communication and programming *(i)* the PLC *(ii)* the screen and *(iii)* the computer with the programming software are connected in the same network to program these devices together.

Figure 5 shows the programmed HMI screens in the Touch Panel, thus obtaining flexibility and ease of use.

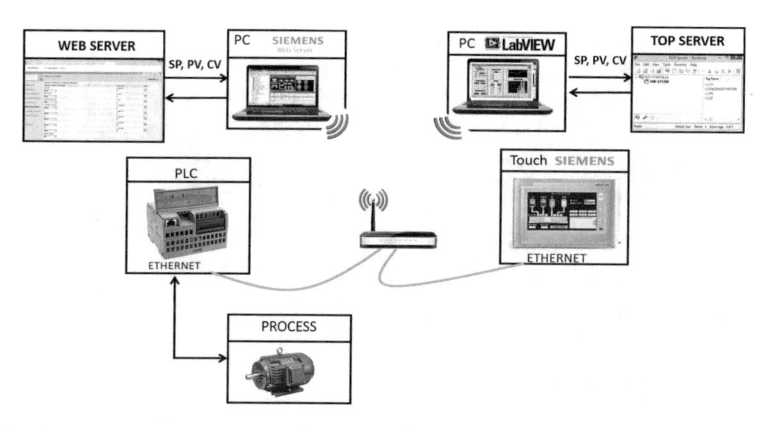

Fig. 4. Interconnection of the HMI for the monitoring and control of the process.

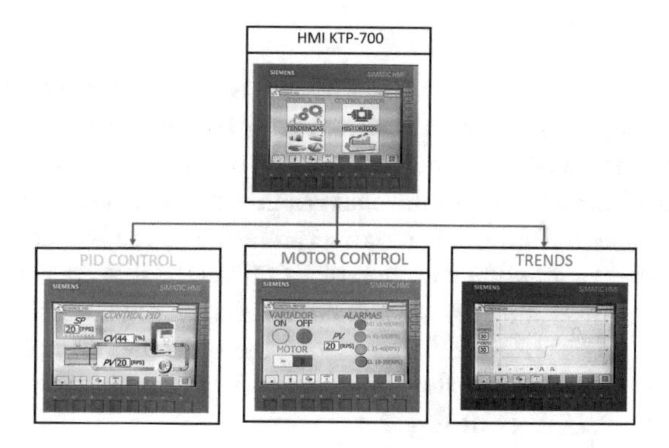

Fig. 5. Screens available on the touch panel HMI.

3.2 Web Server Interface

The range of PLCs of the S7-1200 series in the SIEMENS brand has a webserver that provides access to the variables defined in the CPU of the PLC, thus enabling the implementation of a basic SCADA system Fig. 5. Once the Web Server option is activated in the PLC to later enter the IP address of the PLC in any web browser, we have remote access to the process variables for PID speed control, where we can perform the functions of Monitoring and control in a single interface provided by the web server platform as indicated in Fig. 6.

SIEMENS S7-1200 station_1 / PLC_1

Nombre de usuario
Iniciar

Estado de variables

Indique aquí la dirección de la variable que desea observar

- Página inicial
- Identificación
- Búfer de diagnóstico
- Información del módulo
- Comunicación
- **Estado de variables**
- Navegador de archivos
- Páginas de usuario
- Introducción

Dirección / Formato de visualización	Valor / Valor
MOTOR BOOL	true
SET POINT REAL	30.0
PROCESS VALUE DEC	30
CONTROL VALUE REAL	54.56452
L BOOL	true
LL BOOL	false
H BOOL	false
HH BOOL	false
Nueva variable BIN	

Fig. 6. Web platform of the process.

3.3 LabVIEW Interface

The LabVIEW System Design Software software is used as a resource to perform the HMI interface through a PC, thus allowing the programming of an environment that helps to visualize the values that are generated in the process, they are presented In a graph and help the interpretation of the changes that are generated. The HMI allows the user to insert the value in RPS of the motor as preferred by the operator (SP) and sent to the PLC, all this makes possible with the use of OPC [15] that creates link to send and receive information from the PLC to LabVIEW or vice versa wirelessly as shown in Fig. 4 *(i)* the PLC *(ii)* TOP SERVER and *(iii)* the computer with LabVIEW System Design Software connected on the same network. Following are the screens programmed in the LabVIEW HMI Fig. 7.

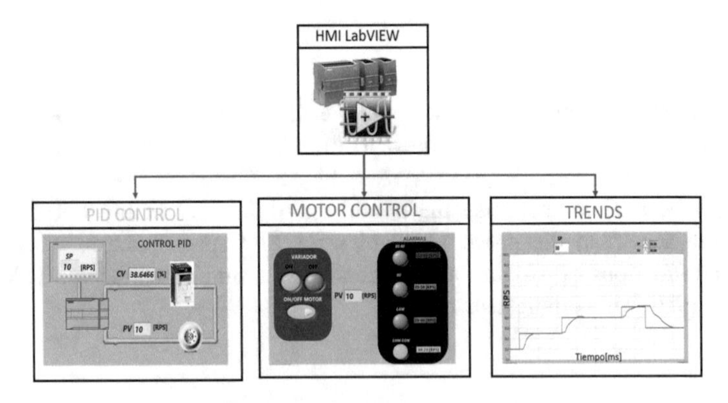

Fig. 7. Screens available in the LabVIEW HMI.

4 Results

For the analysis of the results, the fully constructed module was used, which performs a PID control of a three-phase motor. Figure 8 indicates the module which consists of: *(i)* a Siemens S7-1200 PLC, *(ii)* a Touch Panel model KTP 700, *(iii)* there is a 24Vdc LOGO source which feeds The Touch Panel and the PLC, *(iv)* is has the anlogic output module to generate the control variable (CV); *(v)* to manipulate the motor speed uses a frequency inverter which is connected to the motor.

Fig. 8. Speed control module of a three-phase motor.

The implemented interfaces indicate each of them the state of the process in real time, as shown in Fig. 9, indicating the operation of the PID control on the HMI unrolled in the Touch panel on the screen corresponding to Trends, even on this screen It is possible to make SP changes and check the operation of the PID control used, as you can see, the PID responds correctly before SP changes, stabilizing the PRS in the desired value.

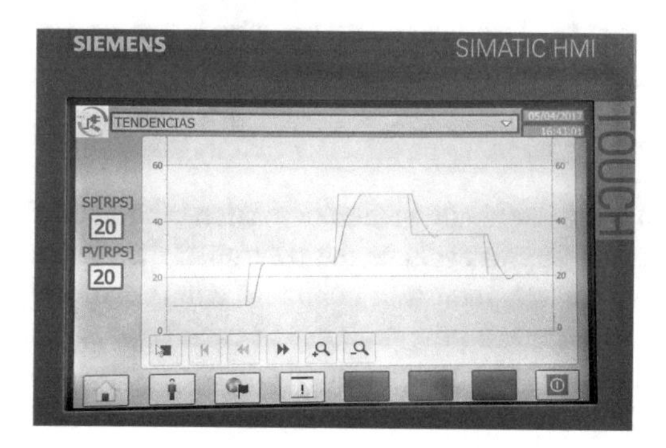

Fig. 9. PID operation shown on the touch panel HMI.

The HMI developed in LabVIEW likewise indicates the trends of the real-time process in which you can visualize the necessary damages that you want to know about the process as shown in Fig. 10 the corresponding tests are performed to verify the control and monitoring.

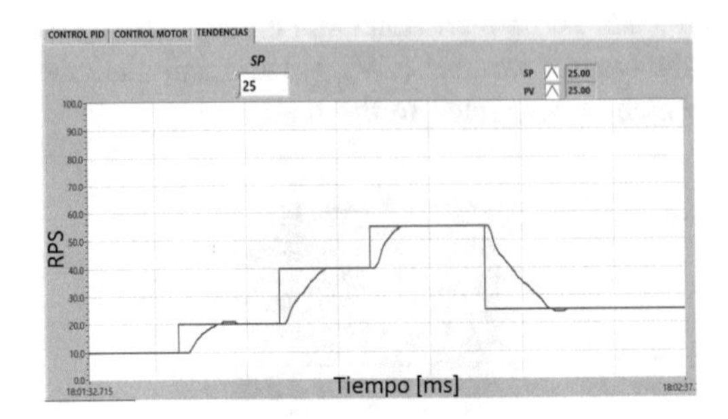

Fig. 10. PID operation shown in the HMI developed in LabVIEW.

The HMI developed in WEB only indicates the values of the process, which are in real time in the same way as the other interfaces, in Fig. 11 the display of the state of the process values is indicated in the same way change the SP as required by the user.

Estado de variables

Indique aquí la dirección de la variable que desea observar

Dirección / Formato de visualización	Valor / Valor
MOTOR	true
BOOL	
SET POINT	25.0
REAL	
PROCESS VALUE	25
DEC	
CONTROL VALUE	57.02763
REAL	
L	true
BOOL	
LL	false
BOOL	
H	false
BOOL	
HH	false
BOOL	

Fig. 11. Status of process variables in web service.

5 Conclusions

The construction of the module allows familiarization with industrial speed processes, as well as in a practical way to evaluate the controller, the module presents industrial solutions for process control, industrial communications and design of HMI screens using different software for the programming of each Of them facilitating the monitoring directly or remotely. The process interfaces are practical of the industrial type with the corresponding screens for monitoring and control of the variables involved in the speed process.

As a future work can be done a graphical interface with Web Server programming in html language to create different screens similar to the messages in the article.

References

1. Balsa-Canto, E.: Algoritmos eficientes para la optimización dinámica de procesos distribuidos, Hdl.handle.net (2017). http://hdl.handle.net/10261/29829. Accessed 07 Apr 2017
2. Greer, A., Newhook, P., Sutherland, G.: Human–machine interface for robotic surgery and stereotaxy. IEEE/ASME Trans. Mechatron. **13**(3), 355–361 (2008)
3. Downs, J.J., Vogel, E.F.: A plant-wide industrial process control problem. Comput. Chem. Eng. **17**(3), 245–255 (1993)
4. Ogata, K.: Ingenieria de Control Moderno, 4ta edn., p. 984. Pearson Educacion, Madrid (2004)
5. Johnson, M., Moradi, M.: PID Control, pp. 539–541. Verlang, Londres (2005)
6. Pruna, E., Andaluz, V., Proano, L., Carvajal, C., Escobar, I., Pilatasig, M.: Construction and analysis of PID, fuzzy and predictive controllers in flow system. In: 2016 IEEE International Conference on Automatica (ICA-ACCA) (2016)
7. Albán, O.V.: control predictivo de un robot tipo scara (2017)
8. Gómez Ortega, J.: Navegación en robots móviles basada en técnicas de control predictivo neuronal, Hdl.handle.net (2017). http://hdl.handle.net/11441/15262. Accessed 07 Apr 2017
9. Mylvaganam, S., Waerstad, H., Cortvriendt, L.: From sensor to web using PLC with embedded web server for remote monitoring of processes. In: Proceedings of IEEE Sensors 2003 (IEEE Cat. No. 03CH37498)
10. Adamo, F., Attivissimo, F., Cavone, G., Giaquinto, N.: SCADA/HMI systems in advanced educational courses. IEEE Trans. Instrum. Meas. **56**(1), 4–10 (2007)
11. Lew, R., Boring, R., Ulrich, T.: A prototyping environment for research on human-machine interfaces in process control use of Microsoft WPF for microworld and distributed control system development. In: 2014 7th International Symposium on Resilient Control Systems (ISRCS) (2014)
12. Shahid, M., Shahzad, M., Bukhari, S., Abasi, M.: Autonomous vehicle using GPS and magnetometer with HMI on LabVIEW. In: 2016 Asia-Pacific Conference on Intelligent Robot Systems (ACIRS) (2016)
13. Azam, M., Khan, K.: Design of the Ethernet based process data extraction algorithm and storage technique for industrial HMI systems. In: 2010 The 2nd International Conference on Computer and Automation Engineering (ICCAE) (2010)
14. Pruna, E., Chang, O., Jimenez, D., Perez, A., Avila, G., Escobar, I., Constante, P., Gordon, A.: Building a training module for modern control. In: 2015 CHILEAN Conference on Electrical, Electronics Engineering, Information and Communication Technologies (CHILECON) (2015)
15. Adell, E., Várhelyi, A., Alonso, M., Plaza, J.: Developing human–machine interaction components for a driver assistance system for safe speed and safe distance. IET Intell. Transp. Syst. **2**(1), 1 (2008)

Virtual Reality on e-Tourism

Juan C. Castro[1(✉)], Mauricio Quisimalin[1(✉)], Víctor H. Córdova[1(✉)], Washington X. Quevedo[1,2(✉)], Cristian Gallardo[1(✉)], Jaime Santana[1(✉)], and Víctor H. Andaluz[2(✉)]

[1] Universidad Técnica de Ambato, Ambato, Ecuador
{juanccastro, hernanmquisimalin, victorhcordova}@uta.edu.ec, cmgallardop@gmail.com, jaimesantanal@hotmail.com

[2] Universidad de Las Fuerzas Armadas ESPE, Sangolquí, Ecuador
{wjquevedo, vhandaluzl}@espe.edu.ec

Abstract. This article proposes the develop of a dynamic virtual environment that with the consumption of real time data about the state of a place, offer an immersion to the tourist like be at the desired location. The development implements a communication and loader structure from many information sources, manual information data loaded from mobile devices and data loader from collecting equipment that get environmental and atmospheric data. The virtual reality application use Google Maps, and worldwide heightmap to get 3D geographic map models; HTC VIVE and Oculus SDK for support virtual reality experience; and weather API to show the weather information from the desired location in real time. In addition, the proposed virtual reality application emphasizes user interaction on the virtual environment by displaying dynamic and up-to-date information about tourism services.

Keywords: Virtual reality · Mobile applications · Tourist experience · Experience quality

1 Introduction

At the international travel and tourism conference of the United Nations, tourism is defined as: *The consumption, production and distribution of services for travelers who dwell in some place other than their domiciles or workplace for at least twenty-four hours. Shorter sojourns are regarded as mere excursions.* On the other hand, tourism without creativity could not survive, that is why it must innovate its forms of diffusion, for which the technological advances cannot be excluded [3, 4].

Any type of technological approach that we want to implement, we must accept that the world has changed, in the global village that we are, we stop talking about isolated societies, and we start talking about figures such as virtual groups. Currently, with the development of new technologies, different virtual groups have been created: *e-health* Which based on technology, takes care of the human's health; *e-business* Uses information and communication to facilitate transactions between businesses and customers; *e-learning* Uses technologies for teaching; *e-government* Is an ICT application for the benefit of e-society, focused on digitizing services, information or

K.J. Kim et al. (eds.), *IT Convergence and Security 2017*,
Lecture Notes in Electrical Engineering 450,
DOI 10.1007/978-981-10-6454-8_13

transactions offered to citizens, or the exchange of information between governmental groups; *e-democracy* Is intimately linked with *e-voting*, *e-participation* and *e-inclusion,* which focuses on democratic procedures [5]. As described above, technology is reaching all areas that humans developed, and for all of this, there is no doubt that e-tourism should not be left out [5].

E-Tourism was born as a society of travelers who exchange experiences of their travels and recommendations of tourist destinations. The internet is the main resource for obtaining information about services related to tourism; These resources can be created and maintained by a web editor, a community or members of a forum [5, 6]. The dissemination of tourist information is born from the desire to have a small experience before taking a decision, traditionally it is done by media content, product presentation, illustrations, catalogs, maps, and other materials [2]. New technologies offer new tools for transmitting information from many places regardless of time or distance [7].

Virtual Reality, VR, focused on tourism, is a technology tool that is currently under development, this allows the tourist to have a closer experience of the desired place, either for fun, distraction or professional purposes [3]. Collaborative Virtual Environments, CVEs, are virtual environments shared by users for their interaction and collaboration, these can be on the same site or can be scattered around the world. This type of virtual reality can be seen in the areas of e-tourism and VR-Shops, in addition to information deployed within virtual environments, integrates the use of real-time data collection ESRI ArcGIS attributes [7].

In Ecuador, tourism is considered as one of the most important economic sources for the country as it is currently the third economic source of non-oil revenues for the Ecuadorian economy. Therefore, with the objective of increasing the source of economic income, the Ministry of Tourism has implemented five strategic pillars at the National level to convert Ecuador into Tourism Power; The strategic pillars focus on: *(i) security* to build trust; *(ii) quality* to generate loyalty; *(iii) destinations and products* to generate unique experience; *(iv) connectivity* to generate efficiency; And finally *(v) promotion* to generate demand. Specifically, in order to comply with the promotion strategy, Ecuador has invested large amounts of money in order to increase the number of foreign tourists visiting the country; As an example is the tourist promotion of the *ALL YOU NEED IS ECUADOR* campaign that emphasizes that no country in the world can say that it has everything like no other, and everything in one place and as close as Ecuador has. For the positioning of the campaign, Ecuador was the first foreign country to promote itself as a tourist destination through a sports event like "Super Bow 2015" with an investment of 3.8 million dollars in a commercial of 30 s plus the cost of production that surpasses the million dollars.

As described in previous paragraphs, this paper presents the development of a virtual touristic application to promote and spread the tourist sites of the Republic of Ecuador, Tungurahua Canton. A flexible application that adapts to situations where variables obtained from dynamic environments are handled, giving the user the opportunity to interact with the environment and adapt it to reality. In contrast to traditional implementations in which virtual environments contemplate the use of static components such as land, buildings, roads, and perhaps some information of its components, climatic conditions, vehicular flow; The proposed application integrates the use of real-time information, both for the dynamism of the scenarios and for the visualization of solids and information, by providing dynamic environments and the ability to read data in real time, the application is completed with integration Of interaction modules so that the user can work in the virtual world according to their needs. In addition, a real-time communication and data loading structure is implemented from different sources of information, either manually from mobile devices, as well as data loading from atmospheric environmental data collecting equipment.

2 System Features

In this research handled data that are collected through an application of information load, furthermore the consumption of APIs of mapping and reception of atmospheric data, all this information is stored in a centralized database the Which standardizes its access and management through API REST technology, making possible the consumption of up-to-date information on the proposed virtual reality application.

A virtual reality application that allows dynamic scenarios and user interaction, must have access to information in real time, to assist in the process of gathering information, in this proposal is developed an application supported on Android Operating System 4.2.2+ that allows to load the information from the desired sites. This information is synchronized to an already defined, created and hosted database on a server with public IP that allow the connection from any user with internet connection; Security politics are integrated against loss of connection with the central server, in that situation, the information is stored in a SQLite database on the mobile device until it can reconnect with the central server and perform a data update (Master-Slave replication). The centralized database is fed by data loader mobile application, but it allows to be fed with information of meteorological centers too. In province of Tungurahua are currently active 10 meteorological stations by the Ecuadorian meteorological service [9].

Although the acquired information corresponds to various types of data, variables and relationships, all information are available for consumption under a REST API schema, so it is not limit the consumption of information to certain devices or development environments, any application on any equipment that has REST API service consumption support can interact with the information repository, see Fig. 1.

In the process of information consumption, the virtual reality application uses the information of the centralized database through HTTP requests, the requests specify the format in which the information will be returned, in this case it uses JSON format, as it allows a high processing speed and generates smaller files that are transmitted at high speed in the network to provide a channel with information in real time.

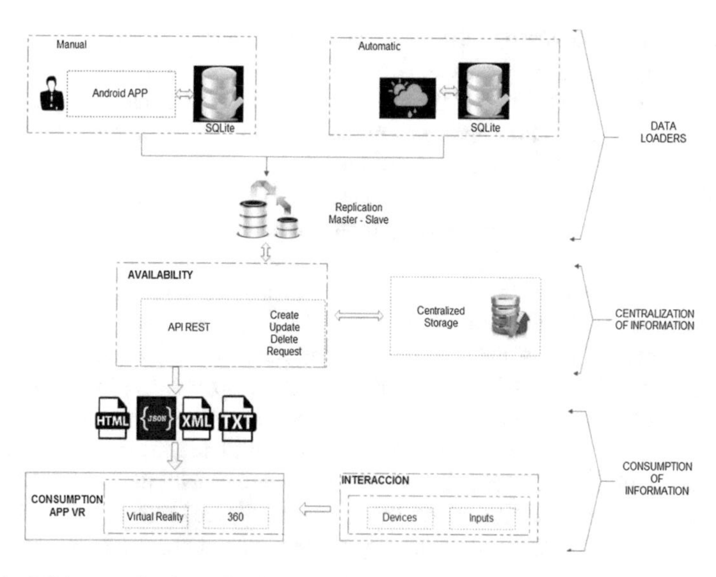

Fig. 1. Block Diagram for the collection of information to be used in real time, by the e-tourism application.

3 System Structure

The control scripts are developed, to respond the operation of different process associated to the virtual device in the Unity environment, see Fig. 2.

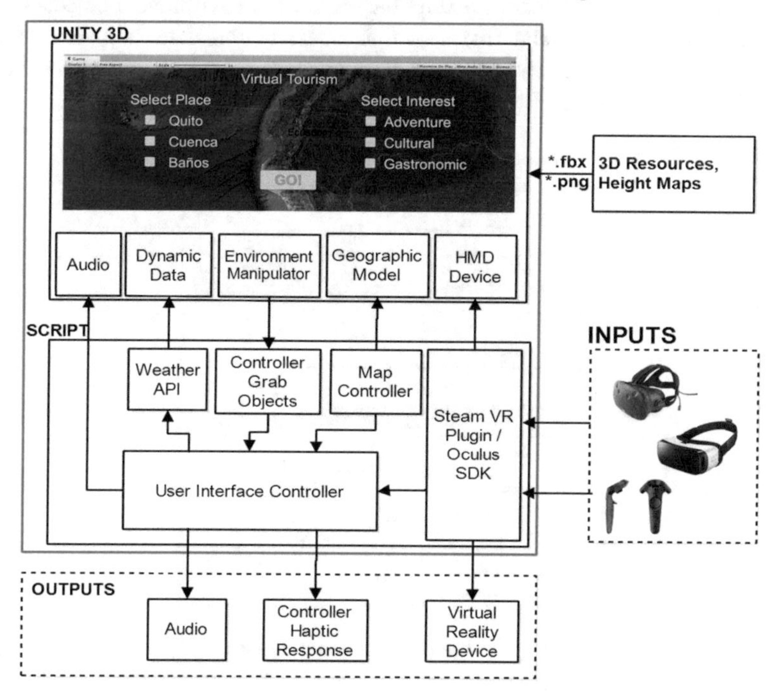

Fig. 2. Component interrelation diagram

The map recreation phase of the scene, contains all virtual reality resources that will be used in the app like Audio effects, show data panels, region maps in 3D, 360 media content and the Game Objects that the virtual Reality Device needs. The goal is to show the resources and information updated immersive, and with a natural input method.

In the phase of inputs and outputs, we have the virtual reality devices (tethered and mobile) that will show the immersive part of the development, providing haptic responses to the interaction and sending input information in the user interface and in the Motion tracking of the HMD.

In the SCRIPTS phase, there are controllers that independently perform their work but are controlled by the User Interface Controller which manages the input forms and generates the information of the output phase and manages the contents in 360 that will be shown according to the behavior algorithm of the App and user interactions. As part of the individual drivers are the API Weather that reads the weather information in the region for example from AccuWeather servers, while the Grab Objects Controller converts the actions of the inputs into events that interact with the selected 3D objects. Finally, the Map Controller is a script that generates the 3D relief of the original map, depending on the interactions with the app the terrain can change to improve the user experience.

4 Virtual Environment

The app development requires a list of steps managed in layers' mode to achieve a complete app. Each layer explains the important areas that conforms the final app like *(i) Layer 1:* the method used to transform a 2D map to a 3D map with the original features of the terrain; *(ii) Layer 2:* texture the terrain and environment objects with the original colors of land reliefs; *(iii) Layer 3:* consumption data from a centralized API REST system, manage 360 media content and interact with the weather API and auxiliary 3D Resources, see Fig. 3.

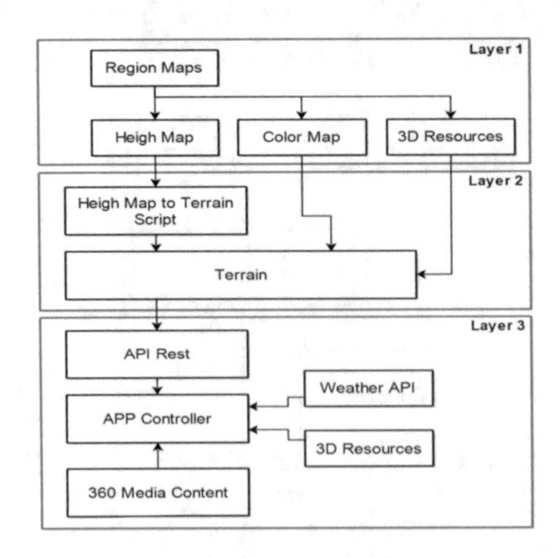

Fig. 3. Layer interaction diagram

The journey stars in the layer creation of real geographical data of the region where these maps are to be worked, the most similar to reality as you can find it in tools like Google Earth, OpenStreetMaps, etc. There are not available real 3D terrain models or a directly way to import a Google Earth 3D map to a Unity scene. A way to solve this problem consist in create a heightmap from a 2D map of the selected region. The Heightmap could be found in the public topography maps of NASA.

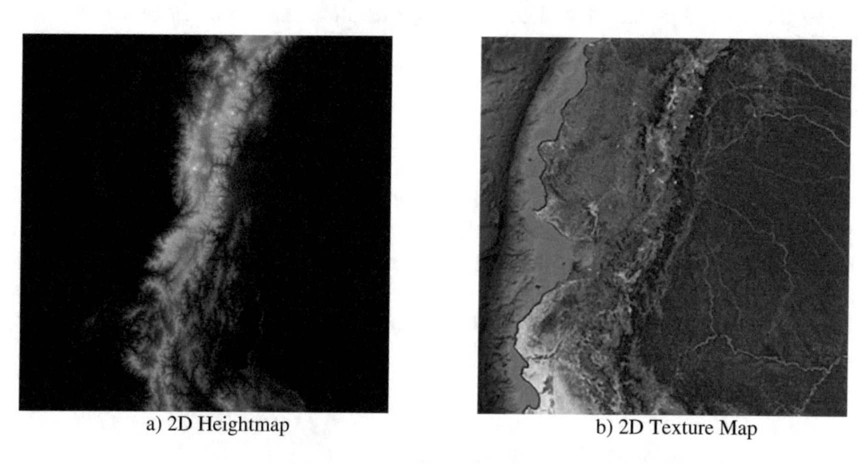

a) 2D Heightmap b) 2D Texture Map

Fig. 4. 2D Maps of Ecuador

A heightmap or heightfield is a raster image used to store values in their grayscale colors, such as surface elevation data, for display in 3D computer graphics. In this case the heightmap is converted into a 3D mesh data for a relief terrain. This process is made by an algorithm implemented in javascript, see Fig. 4. This process is done one time, but as the user interacts or changes the map, the algorithm is executed again to optimize the user experience. The color 2D map is used to texture the 3D map with the original features of the relief, see Fig. 5.

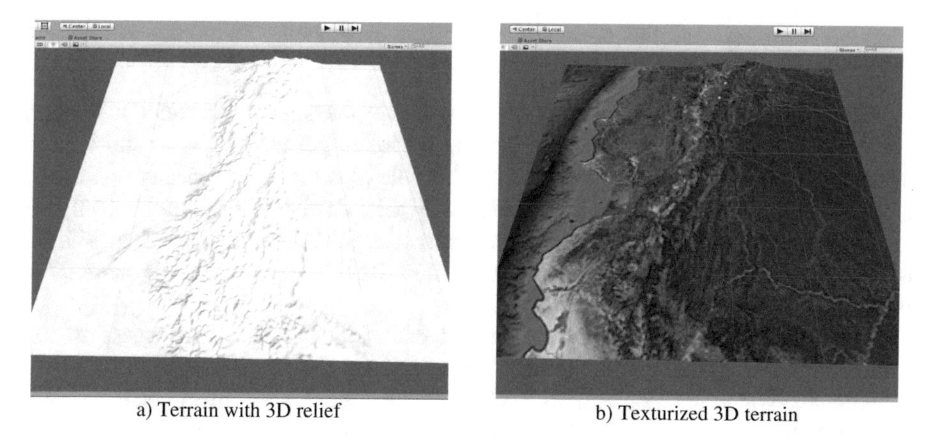

a) Terrain with 3D relief b) Texturized 3D terrain

Fig. 5. 3D terrain model of Ecuador

To share data in the API REST is necessary collect important data from the selected places such as: coordinates, routes, tourist guide, and tourist services. The App Controller access to API REST with the input information from the user to show relevant information. Another activity is managing 360 media like play/pause 360 videos based on a sphere texture player and an audio playback synchronized with the video. The other 3D resources are used like controls of video player, displaying photos, markers of selected places and panels that shows short description that come from a query to the API REST, see Fig. 7.

5 Results and Discussion

This section shows an inmersive virtual environment of Baños de agua Santa, a turistic city of Ecuador. In this case the user could be explore the zone in a real 3D terrain generated with real data from google earth.

Before use the app is necessary to share data in the API REST with a small tool developed specific to this case called The Android Data Loader (ADL), see Fig. 6.

Fig. 6. Android data loader application.

Data loader Android application, help with the proccess of information loading, using the internal gps features of the mobile device to get the longitude and latitude taking care with the best location estimate, this data plus the input capacity to write de aditional information are saved in the internal SQLite database, then automatically the application test the internet connection status to sync the internal SQLite data with the centralized database over internet through API REST features on the centralized database, see Fig. 7.

The centralized base has all the loaded data with the ADL, to access to this information the app query with the notation specified in the API REST, see Fig. 7. The important reason to develop an API REST is that any device or software can access to this information. And now the TourismVR app can run and consume loaded data.

Fig. 7. Centralized database over internet and API Rest Access

When the user opens TourismVR, there are an animation of earth leaving to Ecuador and shows an interface to select place. Quito, Cuenca and Baños and select the tourism type interests: adventure, cultural and gastronomic, see Fig. 8. This input data help to show relevant information. And the next step is press the GO! button to travel to Baños de Agua Santa city.

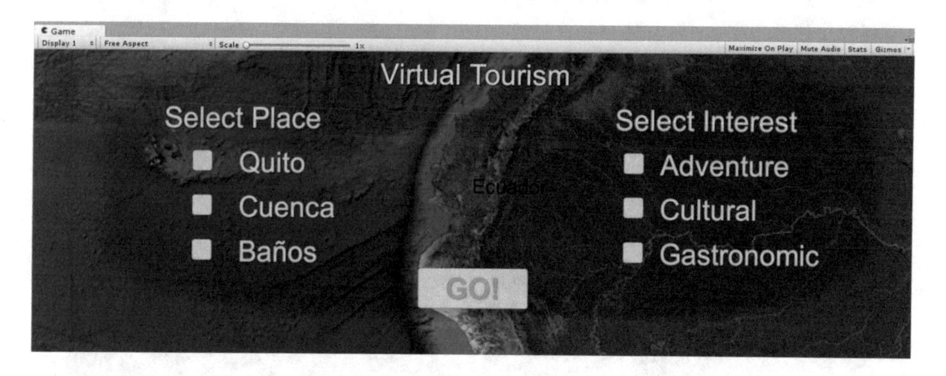

Fig. 8. Home screen TourismVR app

For this prototype app the data for these places is: hotels, restaurants, and touristic activities of Baños de Agua Santa city located in the Tungurahua Province. The experimentation was developed with HTC VIVE headset in a room scale configuration with a Virtual Reality Ready PC.

When the user chooses the site and type of tourism the app carries the user to de top of Tungurahua Volcano that offers a panoramic view of the city and the markers with the suggested places to visit, see Fig. 9. This feature makes the user appreciate the type of nature that have in their hands like: valley, plains and mountains, etc.

Fig. 9. Landing in the top of Tungurahua Volcano

The suggested places are: Ines Maria Waterfalls, Palomino Flores Park and La Virgen Waterfalls. To go to this places, the user interface allows you to scroll inside the environment using controlled flights to each one of suggested places using the grip button of the HTC VIVE's controllers. Another method to go to suggested places is through a guided tour using animations of tarabita, bus, and teletransport directly to the suggested place, see Fig. 10.

Fig. 10. Controller flight around the suggested places

The user could take advantage of this environment to look up the landscape, in this case the user can see a way to transport lava and mud from Tungurahua Volcano when it is active. This dangerous way is close to the city. This natural way cannot see always from the real site in Baños because the weather does not allow it and the difficult to arrive.

When the user is in a suggested place, he will can see the resources, 360 media available related with the place like name and a short description, weather info and the forecast to the next week. In this case the selected place is the La Virgen Waterfalls, see Fig. 11.

Fig. 11. Controller flight around the suggested places

At this point when the user selects the video sphere, the headset shows a 360° video, and when the photography sphere is selected, the headset displays a 360° photos reel from different angle of place. The user interface has buttons to forward or rewind to navigate on 360 media content. The projection of 360 media content give to the user the immersive sensation in the real place because there are real video and photos from the place, see Fig. 12.

a) 360 Photo b) 360 Video

Fig. 12. Displaying 360 media content

At the end, the user has experienced two virtual experiences in different ways. The first one, to navigate in a virtual environment faithful to the reality of the tourist site, and second, to enjoy content in photos and videos 360. The two ways that the app offers to visit a place makes the user feel immersed in the place even before visiting it. The application is in development to load to its bases the information of each tourist place that is wanted to promote.

6 Conclusions

Immersion in virtual reality environments is complete when the interaction between 3D resources are manipulables and this creates empathy between the user and the virtual experience. The main objective is the visualization of informative data about virtual environments, data is static or have a few changes in the course of time, as data obtained in real time, the same ones that help the tourist to know the current state of a desired environment. Finally, the distribution is done for both mobile applications and devices (HTC VIVE and Gear VR), which help potential tourists to know a part of the benefits offered by tourist places. As a prototype of information feed, data are taken from Baños de Agua Santa city, located in the province of Tungurahua - Ecuador.

The use of virtual reality is not a way of replacing traditional tourism, this publication focuses on the dissemination of information so that the tourist has information on the conditions of the site that he wants to visit. Further, the platform made up of centralized data and access through API REST as well as the App Controller, present scalable features to be able to add more information and features as well as services and availability of resources that will give to the user an immersive experience ever greater.

Acknowledgments. The authors would like to thanks to the Technical University of Ambato and through it to the Faculty of Administration for financing the project: Determinants of the tourist enterprise in the cantons of the province of Tungurahua (PFCA04).

References

1. Nee, A.Y.C., Ong, S.K., Chryssolouris, G., Mourtzis, D.: Augmented reality applications in design and manufacturing. CIRP Ann. Manuf. Technol. **61**(2), 657–679 (2012)
2. Reinhart, G., Patron, C.: Integrating augmented reality in the assembly domain-fundamentals, benefits and applications. CIRP Ann. Manuf. Technol. **52**(1), 5–8 (2003)
3. Putro, H.T.: Immersive Virtual Reality for Tourism and Creative Industry Development (2015)
4. Wang, Y., Yu, Q., Fesenmaier, D.R.: Defining the Virtual Tourist Community: Implications for Tourism Marketing (2002)
5. Magoulas, G.D., Lepouras, G., Vassilakis, C.: Virtual Reality in the e-Society (2007)
6. Berger, H., Dittenbach, M., Merkl, D., Bogdanovych, A., Simoff, S., Sierra, C.: Opening new dimensions for e-Tourism. Virtual Real. **11**(2–3), 75–87 (2006)

7. Najafipour, A.A., Heidari, M., Foroozanfar, M.H.: Describing the virtual reality and virtual tourist community (applications and implications for tourism industry). Kuwait Chap. Arab. J. Bus. Manag. Rev. **3**(12), 12–23 (2014)
8. Baldock, P., Burdett, D., Corcoran, P., Stock, C.: Real-time data visualization in Collaborative Virtual Environments for emergency response (2009)
9. Instituto nacional de Meteorología e Hidrología, red de estaciones meteorológicas. http://www.serviciometeorologico.gob.ec/red-de-estaciones-meteorologicas/

Proposing a Key Model e-Commerce Towards Digital Economy for Coastal Areas in Indonesia

Bani Pamungkas[1], Siti Rohajawati[2(✉)], Devi Fitrianah[3], Ida Nurhaida[3], and H.H. Wachyu[3]

[1] Political Science, Bakrie University, Jakarta, Indonesia
bani.pamungkas@bakrie.ac.id
[2] Information System, Bakrie University, Jakarta, Indonesia
siti.rohajawati@bakrie.ac.id
[3] Faculty of Computer Science, Mercu Buana University, Jakarta, Indonesia
{devi.fitrianah,ida.nurhaida,wahyuhari}@mercubuana.ac.id

Abstract. In today's digital economy, the government must face challenges the to acquire, prepare, anticipate, and develop policies strategically for leveraging micro and small business (MSB). As a consequence, MSB and home industry should anticipate their business using information and communications technologies (ICT) into e-commerce platform. This paper is to propose a key model towards digital economy using Performance Evaluation Matrix and Analytical Hierarchical Process (AHP) for MSB in coastal area Jakarta, Indonesia. This research-in-progress study assesses transform factors for build business environment using the internet instrument such as e-commerce, e-business, e-media, and e-government. The triangulation method was used into literature studies, in-depth interview, and secondary data. Based on analysis gap we represented the potential factors key into a scheme of AHP. Our analysis shows that many potential regulation and policies are required to develop an environment as a central market means to foster digital economy and learning in organizations.

Keywords: e-Commerce · The digital economy · Coastal areas

1 Introduction

Indonesia has a chance to be the biggest digital economy in South East Asia by 2020. With more than 93.4 million persons of internet users in the world and 71 million persons are using a smartphone, Indonesia will have 1,000 of technopreneurs with business value IDR 10 billion and e-commerce value of IDR 130 billion.

According to government's vision, the Minister for the Economy announced the XIV Economy Package Policy. In digital economy, the government face several challenges to acquire, prepare, anticipate, and develop policies strategically in leveraging the micro and small business including home industries. For that reason, the government needs to deliver President Regulation on E-Commerce which called a Road Map to encourage expansion and improvement of economy activities efficiently and

K.J. Kim et al. (eds.), *IT Convergence and Security 2017*,
Lecture Notes in Electrical Engineering 450,
DOI 10.1007/978-981-10-6454-8_14

globally connected. The objectives are to encourage young generation in creation, innovation, and invention to new economy activities. The Road Map Regulation has eight aspects of adopting E-Commerce (Table 1) as fully type from 'Majalah' ICT [1].

Table 1. Factor key to success based on authors

Factors key to success	Authors
Infrastructure, legislation, and promoting	[8]
Infrastructure, the digital broadcasting, the Internet, e-services and e-government, applications and IT services, hi-tech equipment, and education	[9]
The Internet, IT application, Technological digital divide, strategies plan, business intelligence, and organic network structure	[10]
Ownership of ICT gadgets/tools/services, ICT, and the internet	[11]
Individual, organizational, environmental, technological, and economic factor	[12]
Individual, organizational, technology, market and industry, external support, and government support	[7]
Web sites marketing, mobile technology, and partnerships	[13]
Internet, technologies, complexity of business, investment, organizational resistance, culture, technical skills and IT knowledge, to be suited to SME business and the products/services, awareness, security, linked to the customer-supplier-business partners, the expertise of SME, and e-commerce standards	[14]
Capital, attitude, awareness, knowledge, and skills	[6]
Technological, organizational, environmental and individual factors. Special to Indonesia SMEs, the crucial factors are technology readiness, perceived benefits, and individual factors (innovativeness, IT ability and experience from owners)	[15]
Pressure: *external, environment, competitive* Readiness: *organization, national, industry* Budget: *investment, cost, perceived benefits, resources* People: *IT knowledge, skill, technical power, mindset* Technologies: *infrastructure, internet speed, security, perceived ease of use* Organization: *culture, structure, nature of business, management support and commitment, size and type, owner's awareness, governance*	[16]
Infrastructure, budgeting, and cost of ICTs, and political environment	[17]

The New Era Economy (the Digital Economy, Economy Internet or Economic Net) would require many challenges and changes business on local, national, and even global [2]. Drew [3] suggests adopting the different strategies for e-commerce and the needs of training and support on the industry sector.

According to Indonesia Bureau of Central Statistics [4] stated that MSB continues to increase approximately 1,361,129 whereas 1,328,147 previously in 2012. The Minister of Communications and Information Technology [5] in the headline news revealed that although Indonesia is currently experiencing a slowdown in economic growth in the last five years, but the industry e-commerce even more rapidly. Therefore, Indonesia is optimistic to become largest players of digital economy in Southeast Asia as well as the next backbone of the national economy. Furthermore, this is also in line with the development of the ASEAN Economic Community (AEC) which is potential for a digital state in Southeast Asia's largest economy by 2020.

The dominance of SMEs will come to the basis of e-commerce business. Based on analysis of Ernst and Young, it shows that online business sales value growth increased by 40 percent each year due to 93.4 million Internet users and 71 million users of smartphone devices in Indonesia" [5]. The point of this research is to find and analyze the potential factors key to success and to make the key model of government policies for micro and small business (MSB), especially for the coastal area.

1.1 Coastal Area – Seribu Islands

As an archipelago state, Indonesia has many coastal areas such as Seribu Islands, Mentawai Islands, Maluku Islands, and so on. It will come to be a market potential of the tourism sector and other sectors.

Seribu Islands is a part region of Indonesia capital city, Jakarta. The Greater Jakarta Metropolitan Area in Indonesia is the 2nd largest agglomeration in the world (after Tokyo) with around 30 million inhabitants. However, the island has a considerable potential for development economy, but the disparity of technology infrastructure and the ability to have enough human resources become a gap with the people of Jakarta on the mainland. This region has four biggest islands (Sibira, Tidung, Pramuka, Harapan, etc.) and the MSB growth is dominated by tourism and resorts. Seribu Islands have a great potential for SMEs to take advantage of e-commerce. Tourism potential is widespread in some of the islands have a tourist attraction for both local and foreign countries and its main economic empowerment into industrial centers, home industry, and the modern market.

1.2 Micro and Small Business (MSB) Characteristics

Micro and Small Business (MSB) is different to Small and Medium Enterprise (SMEs). So, we have to classify to Microenterprise. It is a very small business but considered as the backbone of industrial development in the country [6]. In Indonesia, the size of the company has been categorized based on the total number of full-time workers and annual sales turnover. According to Janita and Kian [7] and refer to UU No. 20/2008, Small Enterprise has characteristics (1) Asset - IDR 50 Million to IDR 500 Million excluding (Land and Building); (2) Revenue or Annual Sales Volume - IDR 300 Million to IDR 2,5 Billion; and (3) Employees are 5 to 19 people. Meanwhile, Medium Enterprise (1) Asset – IDR 500 Million to IDR 10 Billion excluding (land and building), (2) Revenue or Annual Sales Volume - IDR 2 Billion to IDR 50 Billion, and (3) Employees are 20 to 99 people. Sometimes, MSB Indonesia is called street vendor (who has a little business with a number of full-time worker and less than five or having a sales turnover).

MSB are most similarly with Malaysian microenterprise that was start-up business with small resources and limited/few capital. Generally, they must provide of income for supporting daily life and family. Next, they must exist by self-employment with limited resources, capital, and technology. These factors are indicated become the inhibitor to achieve the opportunities in enhancing the adoption of IT into their business performance [6].

1.3 The Factors Key to Success

Based on literature systematic review, we summarized the factors key to success in the digital economy, whereas some of the barriers are to switch to success.

2 Research Methods

This research was used triangulation method. The analysis gap was conducted by comparing between literature study, in-depth interview, and secondary data (which collected by the government of Seribu Islands). The keywords e-commerce, digital economy, micro business, SMEs was used for searching journals and paper on Science Direct, Scopus, Emerald, Springer, Proquest, EBSCO, and so on. In-depth interview was done by focus group discussion (FGD) that presented of government, stakeholder, and online business expertise in the forum which facilitated by government city of DKI Jakarta. The first FGD was attended by 20 persons which represent of stakeholder in several departments and unit organization of Bureau of Economic Province DKI Jakarta. The Second in-depth interview was attended by three persons who representative of Seribu Islands. Finally, the third FGD is related to the seminar which more than 35 persons invited by the committee of BeKraft (the independent state commission that focuses on creative economy). This research is preliminary to investigate and identification the factor keys of e-commerce to success towards the digital economy in coastal areas. On this paper, we described the performance evaluation matrix method to gain the list of factor keys and will enhance the results using AHP-QFD for the next research to achieve our goal of the research.

3 Result and Discussion

Based on the in-depth interview, we analyzed problem arising in using ICT to enhance MSB business. The government has done many programs to encourage and facilitated the MSB onto e-commerce environment. Unfortunately, there are gap to face e-commerce environment as following: (1) Develop a collaboration partnership between Business, Government, higher Education (BGE) to support research and development in e-commerce; (2) Promote the MSB to use the big channel of commercial market such as Tokopedia, GoFood, Lazada, so on, and future with Alibaba.com; (3) Give the program training and education to create website, free material, and how to business online transaction; (4) Develop annually the forum discussion to society of Seribu Islands to leverage the potential economy. The biggest problem of this region is transportation and accessing to get the islands. Limited of ships and port become the main barrier and expensive of transportation. Sometimes one ship should pay IDR 11 million for two ways in order to pick and distribute the MSB products;

(5) ICT provider is limited causing led to weak of the signal; (6) Type of MSB are most numerous with a few capital such as anchovies, pickles, crackers, breadfruit. If the material unavailable it causes the business stop operation.

Based on secondary data, we found that BeKraft was initialized to help the problem arising for Indonesian vision to the digital economy. We captured some opinion from Mr. Rudiantara as Ministry of Communication and Information of Indonesia and Mr. Harsono as Chairman Mastel Institute [1]. Herewith the following of their opinion for Indonesia road map e-commerce. The road map exclusively describes on sub-section.

3.1 Indonesia Vision to Digital Economy

Indonesia has 57 million SMEs whereas every year that contributes 58–60% of national GNP. Meanwhile, Government's has an effort to digitalize SMEs or MSB without disruption the current trading activities. Furthermore, some e-commerce expertise was stated in *Majalah* ICT (2016). According to them, Indonesia has a great potential to increase e-commerce industry sectors. Supporting ICT to e-commerce activities will reach out a larger all Indonesia society. They believed that evidence of many innovations will rise digital economy to a better local market opportunity by capital business. This

Table 2. Strategies and point target of Indonesian vision road map (source [1])

Strategies	Point of target
Funding	(1) People Business Credit (KUR) for platform developer tenants; (2) grant for business incubator of startup partner; (3) USO fund to digital UMKM and startup e-commerce platform; (4) angel capital; (5) seed capital from Host Father; (6) crowdfunding; and (7) DNI opening
Tax	(1) Tax redemption for local investors who invest in a startup; (2) simplification of license/tax procedures for e-commerce startup whose profit below RP. 4.8 billion/year; and (3) tax regulations equality for all e-commerce entrepreneurs
Consumer protection	(1) Government Regulation on Trading Transaction using Electronic System; (2) harmony in regulations; (3) payment system for government goods/service trading via e-commerce; and (4) progressive national payment gateway
Education and human resources	(1) E-commerce awareness campaign; (2) national incubator program; (3) e-commerce curriculum; (4) e-commerce education to consumers, doers and law enforcement
Logistic	(1) Utilization of National Logistic System; (2) local courier company enforcement; (3) UMKM logistic data development; (4) logistic development from the village to city
IT infrastructure	Communication infrastructure throughout broadband network construction
Cyber security	(1) National system supervision arrangement in e-commerce transaction; (2) cyber-crime public awareness; (3) SOP arrangement related to consumer data record, certification for consumer data security
Evaluating	Forming executor management by monitoring and evaluating e-commerce road map implementation

condition will support by the government with the various program through collaboration between pioneer company of digital technologies (Google) and program corporate social responsibility (i.e., Pertamina etc.) that will bring positive impacts of business in Indonesia. As in fact, Indonesia has 1.65% entrepreneurs (March 2016), it will become a chance to make Indonesia a host in digital technology. All working cabinet was agreeing to push e-commerce development as one target for Indonesia's digital economy growth.

For this reason, the government has launched the road map to form Indonesia dealing with ecosystem and structure of e-commerce. This aim is to encourage and motivate the business growth as tools to well monitor of using ICT. Meanwhile, the transaction has been reported to increase the value of e-commerce IDR 12 billion (2014) to IDR 18–19 billion (2015). However, the people who are involving in legislative, judicative, and executive should be aware of switching the direct to online interaction. And, create the regulation and law protection on secure and free trade which can diminish tax objects of products and service trading. Indonesia vision Road map to e-commerce is listed in Table 2.

3.2 Propose a Key Model to MSB Success on Digital Economy

Based on results of triangulation method we proposed a key model to success MSB on digital economy that refers to factor keys on SME's critical success or barrier, potential factors of the in-depth interview, and Government road map e-commerce (Fig. 1). We eliminated some factors due to characteristics of coastal area and priorities of strategies implementation.

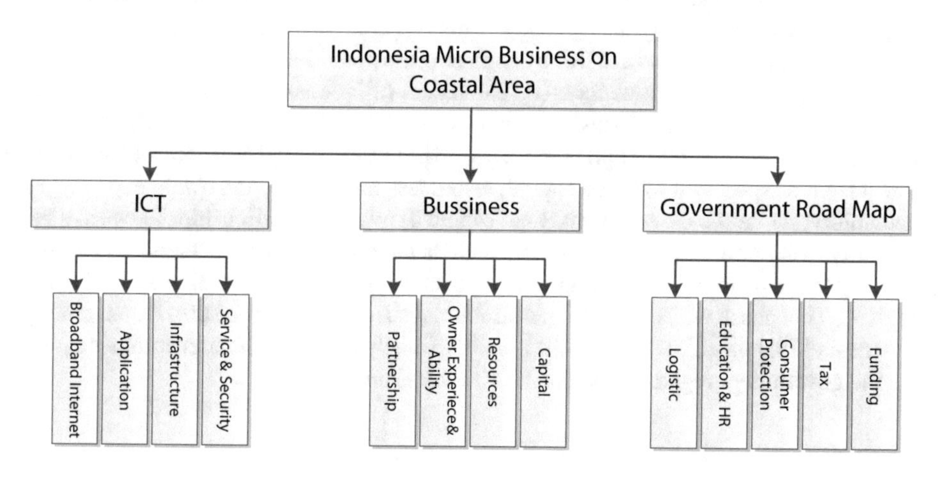

Fig. 1. Factor keys to success using AHP diagram

3.3 Next Step Research

The research will provide insights to the government, researchers, and expertise of e-commerce by establishing the recommendation strategies implementation and adoption

ICT in MSB in the coastal area. The research will determine the priorities factor key that leads to the successful deployment using the Performance Evaluation Matrix with two performance indices, easiness (E), and importance (I) to evaluate the performance factor keys to success on the digital economy [18]. As mention by Chen et al. [18] easiness and importance of factors of implementing e-commerce vary with the amount and distribution of product or material and employee in a company. K scale to assess the easiness and importance will use for each implementation factor. After this method, AHP method will apply to priorities the factor keys. AHP has been widely used in decision-making analysis in various fields such as business, economic, political, social, and management sciences. The research will also focus on the challenges facing the Indonesian government onto the business environment. Again, some SMEs Indonesia studies show the limited of adoption e-commerce for their business sector, while the literature highlights the absence of MSB categories on the take-up rate of e-commerce for coastal areas. This gap have planned and will be examined in the proposed research framework, and consists of three aspects, how the ICT readiness of MSB to play the role of e-commerce environment and structure; what the recommendation strategies policies to lead e-commerce for MSB in coastal areas in order to facilitate them as well as increasing the take-up rate and improvements in the digital economy.

4 Conclusion

Indonesia MSB will play an important role in the digital economy and national economic productivity. Meanwhile, Indonesia vision road map e-commerce was launched to utilize e-commerce and ICT adoption dealing with has the potential to increase the business sectors regionally and globally. This research was proposed a key model to success in digital economy for MSB by exploring the potential factors and seeing insight the Indonesia Road Map e-commerce. The model of factor keys was categorized into three groups, which are ICT, Business, and Government Roadmap. The ICT aspect will investigate Infrastructure, Application, Broadband Internet, and Services and security. The business aspect will focus on capital, resources, partnership, and owner experiences and ability; while the Government Road Map will evaluate relationships between strategically existing and policies planning such as data jurisdiction, regulation, and law. The aspects will focus on funding, tax, consumer protection, education and HR, and logistic. This phase will be based on a structured survey that will be distributed among Indonesia MSB on coastal areas. The results will help to establish the recommendation for the government regarding on policies implementation.

References

1. Majalah ICT: Here it comes: Indonesian e-commerce road map, IVth edn., no. 50. Majalah ICT (2016)
2. Adhikara, C.T.: Siapa konsumen kita?: analisis perubahan konsumen di era “Ekonomi Baru”. The winners **6**(2), 175–183 (2005)

3. Drew, S.: Strategic uses of e-commerce by SMEs in the east of England. Eur. Manag. J. **21**(1), 79–88 (2003)
4. BPS: Perkembangan Data Usaha Mikro, Kecil, Menengah (UMKM) dan Usah Besar (UB) 2012–2013, BPS (2013). https://www.bps.go.id/
5. Kominfo: Indonesia Akan Jadi Pemain Ekonomi Digital Terbesar di Asia Tenggara (2016). https://kominfo.go.id/index.php/content/detail/6441/Indonesia+Akan+Jadi+Pemain+Ekonomi+Digital+Terbesar+di+Asia+Tenggara/0/berita_satker
6. Hairuddin, H., Laila, N., Ab, A.: Why do microenterprise refuse to use information technology \xE2\x80: a case of batik microenterprises in Malaysia, vol. 57, pp. 494–502 (2012)
7. Janita, I., Kian, W.: Barriers of B2B e-business adoption in Indonesian SMEs: a literature analysis. Procedia Comput. Sci. **17**, 571–578 (2013)
8. Pookaiyaudom, S., Samakoses, V.: Three factors key to success of digital economy', forum told (2015). http://www.nationmultimedia.com/news/business/macroeconomics/30253597
9. Mertai, G., Berdykulova, K., Ismagul, A., Sailov, U., Kyzy, S.Y.: The emerging digital economy: case of Kazakhstan. Procedia Soc. Behav. Sci. **109**, 1287–1291 (2014)
10. Foong, L.: 6 key factors influencing B2B marketing in the new digital economy factors B2B marketers need to prepare for in the new digital economy. CEO the ALEA group (2016). http://louisfoong.com/6?key?factors?influencing?b2b?marketing?in?the?new?digital?economy/. Accessed 16 Nov 2016
11. Azman, H., Salman, A., Razak, N.A., Hussin, S., Hasim, M.S., Sidin, S.M.: Determining critical success factors for ICT readiness in a digital economy: a study from user perspective. Adv. Sci. Lett. **21**, 1367–1369 (2015)
12. Consoli, D.: Literature analysis on determinant factors and the impact of ICT in SMEs. Procedia Soc. Behav. Sci. **62**, 93–97 (2012)
13. Kabanda, S., Brown, I.: A structuration analysis of small and medium enterprise (SME) adoption of E-commerce : the case of Tanzania. Telemat. Inform. **34**, 118–132 (2017)
14. Kartiwi, M., MacGregor, R.C.: Electronic commerce adoption barriers in small to medium-sized enterprises (SMEs) in developed and developing countries: a cross-country comparison. J. Electron. Commer. Organ. **5**, 25–51 (2007)
15. Rahayu, R., Day, J.: Determinant factors of e-commerce adoption by SMEs in developing country: evidence from Indonesia. Procedia Soc. Behav. Sci. **195**, 142–150 (2015)
16. Kurnia, S., Choudrie, J., Mahbubur, R., Alzougool, B.: E-commerce technology adoption: a Malaysian grocery SME retail sector study. J. Bus. Res. **68**(9), 1906–1918 (2015)
17. Karanasios, S.S.: An E-commerce Framework for Small Tourism Enterprises in Developing Countries. Victoria University, School of Information Systems, Wellington (2008). Submitted by: Stan Stergios Karanasios Statement of Originality
18. Chen, S.C., Yang, C.C., Lin, W.T., Yeh, T.M., Lin, Y.S.: Construction of key model for knowledge management system using AHP-QFD for semiconductor industry in Taiwan. J. Manuf. Technol. Manag. **18**(5), 576–598 (2007)

A Fixed-Function Rendering Pipeline with Direct Rendering Manager Support

Nakhoon Baek[1,2,3(✉)]

[1] School of Computer Science and Engineering, Kyungpook National University, Daegu 41566, Republic of Korea
oceancru@gmail.com
[2] Software Technology Research Center, Kyungpook National University, Daegu 41566, Republic of Korea
[3] dassomey.com Inc., Daegu 41566, Republic of Korea

Abstract. In the ordinary graphics applications, they use one of the high-level 3D graphics libraries such as OpenGL and Direct3D. From the 3D graphics library implementer's point of view, they use some low-level graphics tools including DRM (direct rendering manager). In this paper, we present the accelerated way of achieving 3D graphics features through directly accessing DRM system calls, rather than using the high-level graphics libraries. With DRM features, we can achieve much accelerated way of typical and simple 3D graphics rendering.

Keywords: Direct rendering manager · Acceleration · 3D graphics output

1 Introduction

In these days, we have plenty of 3D graphics outputs and user interfaces. Most 3D graphics applications use a 3D graphics library or a 3D graphics engine based on the specific 3D graphics library. At this time, OpenGL [1] and DirectX [2] are among the most widely used 3D graphics libraries.

In some graphics applications, the display speed may be one of the most critical factors. Since the data set is too large for most 3D graphics libraries, a little bit of speedup in the rendering process can result in the much speed-ups.

In this paper, we present a new way of accelerated 3D rendering, based on the Direct Rendering Manager (DRM) [3]. For some specific cases, the application programs require only the fixed-function rendering schemes, and then, we can implement the specific functionality into the DRM-based low-level 3D graphics programs, instead of the traditional OpenGL or DirectX. We represent the concepts and designs for the DRM-based accelerated rendering system.

K.J. Kim et al. (eds.), *IT Convergence and Security 2017*,
Lecture Notes in Electrical Engineering 450,
DOI 10.1007/978-981-10-6454-8_15

2 Design of the Rendering System

The Direct Rendering Manager (DRM) is actually a module of the Linux kernel. Its major role is to provide an Application Programmer's Interface (API) to the Graphics Processing Unit (GPU). A programmer can send the rendering commands and the target data to the GPU, through calling DRM functions, as shown in Fig. 1.

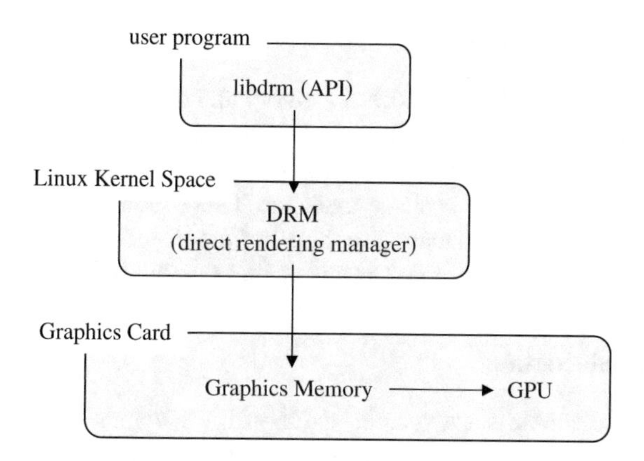

Fig. 1. The DRM module in the Linux kernel.

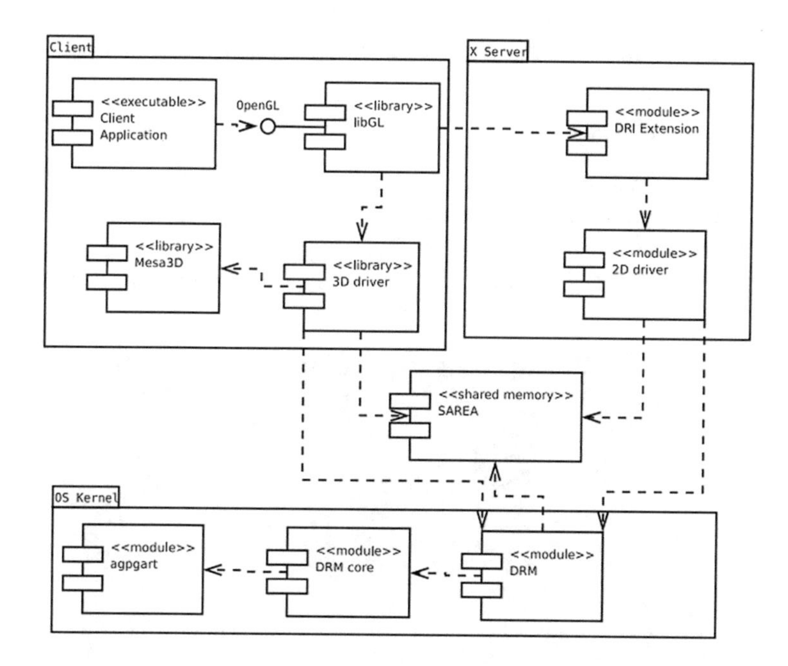

Fig. 2. Direct rendering architectural diagram [4].

The DRM module provides additional functionalities including framebuffer managing, mode setting, memory sharing objects handling, memory synchronization, and others. Some of these expansions had carried out their own specific names, such as Graphics Execution Manager (GEM) or Kernel Mode Setting (KMS). Those parts are actually the sub-modules of the whole DRM module. Figure 2 shows the internal architecture of the DRM module.

2.1 Graphics Execution Manager

Due to the enlarged size of graphics card memory and the complexity of graphics APIs such as OpenGL, the strategy of reinitializing the graphics card state at each context switch was too expensive. Also, Linux kernels need more optimized ways to share graphics buffers with the compositing manager. These requirements led to the development of new methods to manage graphics buffers inside the kernel. The Graphics Execution Manager (GEM) emerged as one of these methods [4].

2.2 Kernel Mode Setting

The graphics card should set a specific mode of the graphics display. So, the screen resolution, number of bits for the colour and depth representation, refresh rate, and others are set by this mode. This operation usually requires low-level access to the graphics hardware. A mode-setting operation must be performed prior to start using the framebuffer, and also when the mode is required to change by an application or the user.

2.3 KMS Device Model

From an existing OpenGL application program, we can extract the shader programs for a fixed rendering pipeline. Then, we use the low-level DRM functions to build-up a minimum system to execute the extracted shader programs. In this case, our application level interfaces are directly connected to the Linux kernel level support to the GPU

Fig. 3. An example 3D graphics output from our DRM-based system.

executions. Based on this simple and intuitive idea, we have implemented the first prototype system, as shown in Fig. 3. This system can display 3D graphics primitives without any commercial graphics library, such as OpenGL and DirectX.

3 Conclusion

In these days, most graphics applications use the commercial implementations of 3D graphics libraries or 3D graphics engines. In contrast, we represent yet another rendering system, based on the DRM. Since DRM is a low-level graphics module at the Linux kernel level, it is hard to achieve a full set of high-level graphics operations. We aimed to provide simple and intuitive fixed-function rendering functions, and provides the design of our system. We expect that this system can be a solution for time-critical large-scale data visualization applications.

Acknowledgements. This research was supported by Basic Science Research Program through the National Research Foundation of Korea (NRF) funded by the Ministry of Education, Science and Technology (Grant 2016R1D1A3B03935488).

References

1. Segal, M., Akeley, K.: The OpenGL Graphics System: A Specification, Version 4.5 (Core Profile), Khronos Group (2016)
2. Luna, F.: Introduction to 3D Game Programming with DirectX 12, Mercury Learning and Information (2016)
3. Faith, R.E.: The Direct Rendering Manager: Kernel Support for the Direct Rendering Infrastructure (2016). http://dri.sourceforge.net/doc/drm_low_level.html
4. Fonseca, J.: Direct rendering infrastructure: Architecture (2005)

Internet of Things

Implementation of an Adaptive Design for the Iterative-MIMO Smart Sensor Detectors to Increase Energy Efficiency in Realistic Channel Conditions

Nina Tadza(✉)

Universiti Tun Hussein Onn Malaysia, Beg Berkunci 101,
Parit Raja, 86400 Batu Pahat, Johor, Malaysia
nina_tadza@yahoo.com
http://www.uthm.edu.my

Abstract. This paper investigates the adaptivity of the Fixed Sphere Decoder (FSD) algorithm, for iterative-multiple-input multiple-output (MIMO) detection in 4G LTE environment. The switching mechanism for the FSD depends on the calculated mutual information between the transmitters and receivers in real-time. The detector determines whether the receiver would detect the incoming symbols using a higher accuracy detector, a less performance detector or simply abandon further processing and reduce energy consumption by requesting a re-transmission. This paper provides the performance analysis for the proposed algorithm in realistic conditions by providing a detailed energy analysis of the algorithm for spatially correlated channel conditions. Analytical, simulation and implementation results show that the practical behavior of the proposed Iterative-MIMO detector saves significant energy with a tolerable bit error rate performance degradation.

Keywords: Fixed sphere decoding · Soft-decoding · Energy savings · Iterative-MIMO · Mutual information · Adaptive Switching Algorithm · FPGA · Spatial correlation

1 Introduction

To meet the explosive growth in data rates currently caused by mobile devices such as smart phones and portable handheld multimedia devices, as well as data terminals such as smart sensors, wireless hotspots, femtocells and base stations, the technology of utilizing multiple antennas on both sides of the transmitter and receiver is imperative. Theoretical analysis has shown promising capacity growth by employing the multiple-input multiple output (MIMO) scheme [1, 2], which helps in increasing the spatial diversity and capacity of the system. However, the presence of spatial correlation between the multiple antennas reduces the capacity improvement [3]. Studies

The research is in conjunction with University Tun Hussein Onn Malaysia, Johor, Malaysia.

K.J. Kim et al. (eds.), *IT Convergence and Security 2017*,
Lecture Notes in Electrical Engineering 450,
DOI 10.1007/978-981-10-6454-8_16

have evaluated the behavior of detectors in such spatially correlated channel environments, for both low complexity linear MIMO detectors [4, 5] and high performance tree search detectors [6]. Generally, it is found that the bit-error-rate (BER) degrades as the channel gets more correlated. Studies are lacking however, for adaptive iterative-MIMO detection as well as for a full receiver setup that includes iterative decoding in such channel conditions. Moreover, to the best of the authors' knowledge, the energy analysis of adaptive algorithm implementations is also sparse in the literature. None of these papers considers the performance of such algorithms in spatially correlated channels or the energy savings potential for realistic hardware implementations. In practice, the channels between different antennas are correlated and therefore the full multiantenna gains may not always be obtainable. Therefore, the work investigates the utilization of the Adaptive Switching Algorithm on simulated spatially correlated channels, whereby the information between the antennas, which is the mutual information (MI) is not optimal.

Therefore, this trade-off of complexity and energy savings gained in the detector in spatially correlated channels are made and justified for realistic design implementations for the Adaptive Switching Algorithm smart sensor receivers.

The main contributions of this paper are summarized as follows:

- The proposed adaptive algorithm is found to control the detector; to choose the appropriate detection method, according to the channel conditions to help minimize resources per transmission.
- Energy analysis and hardware design implementation for the Adaptive Switching Algorithm saves energy whilst maintaining the performance of the detector in spatially correlated channels with only a slight increase in hardware utilization complexity, and higher signal-to-noise ratio (SNR).

2 Spatially Correlated MIMO Channel Model

In order to verify the effectiveness of the Adaptive Switching Algorithm in realistic conditions, spatially correlated MIMO channels are chosen as a reasonable model for providing simulated environments mimicking heavily built-up urban transmission settings for radio signals [7, 8]. Based on a flat fading standard MIMO model [9], with M transmitters and N receivers where $M \times N$, the channel setup considered in this paper utilizes the Kronecker model, where the correlation between the transmitters and receivers are assumed to be independent and separable. This model is reasonable when there is a lot of signal scattering that occurs close to the transmitting and receiving antenna arrays. The results of this model has been validated by both outdoor and indoor measurements [10, 11]. In this case, with Rayleigh fading, the channel matrix can be factorized as in Eq. (1).

$$\mathbf{H} = \mathbf{R}_{Rx}^{1/2} \mathbf{H}_{\mathbf{w}} (\mathbf{R}_{Tx}^{1/2})^T \tag{1}$$

The antenna correlation observed at the receiver is assumed to be the same for all transmitters, and similarly, the correlation for the transmitters is also the same on all receivers. The elements of $\mathbf{H}_{\mathbf{w}}$ are independent and identically distributed (i.i.d) as circular symmetric complex Gaussian with zero mean, μ and unit variance, σ with $\mathbf{vec}(\mathbf{H}) \sim CN(\mathbf{0}, \mathbf{1})$ representing the MIMO uncorrelated channel. The $N \times N$ matrix $\mathbf{R}_{Tx}$ describes the fading correlation for the transmitter array while the $M \times M$ matrix $\mathbf{R}_{Rx}$ described the received spatial correlation. The statistical behavior of the channel matrix can also be expressed as in Eq. (2), where the $\mathbf{vec}(\cdot)$ denotes the vec operator and $\otimes$ the Kronecker product [10].

$$\mathbf{vec(H)} \sim CN(0, \mathbf{R}_{Tx} \otimes \mathbf{R}_{Rx}) \tag{2}$$

The spatial correlation depends directly on the eigenvalue distribution of the correlation matrices, $\mathbf{R}_{Tx}$ and $\mathbf{R}_{Rx}$. Each eigenvector represents a spatial direction of the channel and the corresponding eigenvalue describes the average channel and signal gain in a specified direction. High spatial correlation indicated by a large eigenvalue spread in $\mathbf{R}_{Tx}$ or $\mathbf{R}_{Rx}$ means that some spatial directions are statistically stronger than others. Low spatial correlation on the other hand, is represented by a small eigenvalue spread in $\mathbf{R}_{Tx}$ or $\mathbf{R}_{Rx}$, meaning that almost the same signal power can be expected from all spatial directions. The higher the spatial correlation, the more impact it has on the performance of a given MIMO system [12]. The capacity of the channel is always degraded by the receiver side of spatial correlation as it decreases the number of (strong) spatial directions that the signal is received.

The correlation model considered in this paper can be calculated mathematically with respect to capacity, using generic definitions for the transmitter,

$$\mathbf{R}_{Tx} = \begin{pmatrix} 1 & C_{Tx} & \cdots & C_{Tx}^{N-1} \\ C_{Tx} & 1 & \ddots & \vdots \\ \vdots & \ddots & 1 & C_{Tx}^{2} \\ C_{Tx}^{N-1} & \cdots & C_{Tx}^{2} & 1 \end{pmatrix} \tag{3}$$

and receiver correlations.

$$\mathbf{R}_{Rx} = \begin{pmatrix} 1 & C_{Rx} & \cdots & C_{Rx}^{N-1} \\ C_{Rx} & 1 & \ddots & \vdots \\ \vdots & \ddots & 1 & C_{Rx}^{2} \\ C_{Rx}^{N-1} & \cdots & C_{Rx}^{2} & 1 \end{pmatrix} \tag{4}$$

where $\mathbf{C}_{Tx}$ and $\mathbf{C}_{Rx}$ represents real-valued correlation coefficients. The correlation indexes considered are further simplified to give $\mathbf{R}_{Tx} = \mathbf{R}_{Rx} = \mathbf{C}$, yielding a single factor parameter. This means that the system considers the same correlation is present at both transmitter and receiver sides. The given model can range from the uncorrelated case i.e. $\mathbf{C} = 0$ to the fully correlated scenario of $\mathbf{C} = 1$.

Two points should be understood concerning the use of this model in the paper. First, while the channel model does represent close to realistic channel conditions, the results give pessimistic performance predictions for highly correlated fading scenarios where the model assumptions described above are no longer valid [13]. Secondly, though the correlation values between the transmitters and receivers are unlikely to be equal, this assumption is made to give an overall idea of the applicability of the Adaptive Switching Algorithm to spatial correlated channels.

3 Adaptive Switching Algorithm Description

From the authors' work in [14], it can be seen that the Adaptive Switching Algorithm comprises selecting between by two well-known detection algorithms, namely the Fixed Sphere Decoder (FSD) [15] and the Vertical Bell Laboratory Layered Space Time with Zero Forcing (V-BLAST/ZF) [16] detection algorithms according to the BER performance of the system. Switching between the two algorithms is determined by thresholds pre-calculated from the MI between the transmitter and the receiver, according to the real-time channel conditions of each data transmission.

The experiments for the proposed work use a software/hardware setup performed in Matlab™ and its built-in Simulink® package as well as Xilinx® System Generator to compile into a field programmable gate arrays (FPGA). The transmission setup comprises $M = 4$ transmitters and $N = 4$ receivers, based on a bit-interleaved coded modulation (BICM) setup, which has a transmit frame size of $K_u = 1,024$ bits for transmission over a random independent fast fading propagation channel, $\mathbf{H}$, and it is constant over a frame and changes independently from frame to frame following the Kronecker model, which is perfectly known at the receiver. The transmitted bits, K_u, are encoded using an iterative-turbo scheme at rate of $\phi = 1/2$, which are then interleaved randomly to give, K_a coded bits, before mapping into a quadrature amplitude modulation (QAM) constellation, $\emptyset$ of size $W = 16$ points, forming a sequence of $K_s = K_e = \log_2 W$ symbols. This gives $K_s = 512$ symbols, which are divided equally between the transmitters for 100,000 channel realizations. This part of the transmitter system is simulated purely using Matlab™.

The work focuses on the receiver, which is consequently divided into the theoretical software experimentation and the hardware design implementation. On the software side, the Adaptive Switching Algorithm for the iterative-MIMO receiver is designed in Matlab™ and its built-in Simulink™ modeling package. On the other hand, the hardware design involves constructing the circuitry of the receiver using Xilinx® System Generator based on the latest Xilinx® Virtex-7 hardware. The system setup for both software and hardware co-simulation is shown in Fig. 1.

The power readings are initially estimated by the Xilinx® Power Estimator (XPE)™ tool based on the multiplier resource counter utilization during the software modeling portion. The power readings measured gives ballpark estimates for hardware design, which are later confirmed during the implementation using the Xilinx® System Generator using the Xilinx® Power Analyzer (XPA)™ tool after the model is synthesized and mapped onto the appropriate hardware.

It should be noted that, once the channel realizations and each corresponding channel ordering for specific detection algorithms that make up the Adaptive Switching Algorithm are simulated on Matlab™, the model of each detection and decoding method is demonstrated on Simulink® before being synthesized and mapped onto the Xilinx® Virtex-7 using the Xilinx® System Generator. The hardware is run at a core

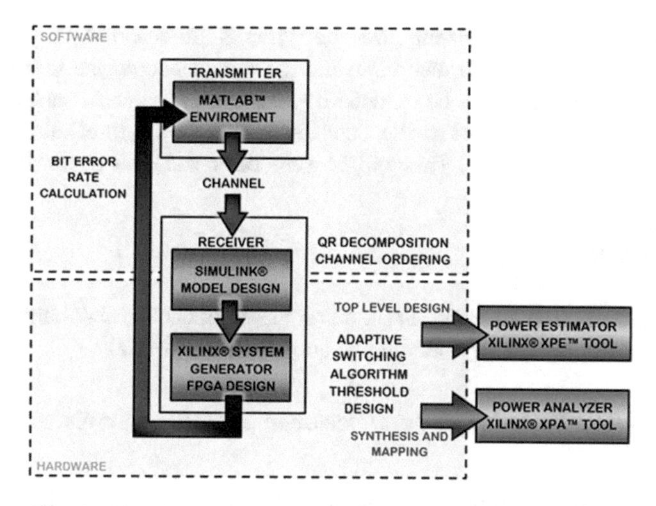

Fig. 1. Flowchart of the software/hardware experimental setup

voltage of 1 V and at the operating frequency of 250 MHz. The energy can then be calculated from the power and the time it takes to transfer and decode a packet. In order to understand how the Adaptive Switching Algorithm receiver is implemented specifically, each block of the FPGA design is described in detail in [17].

3.1 Adaptive Switching Algorithm Implementation

The general idea behind the Adaptive Switching Algorithm is explained in Fig. 2. The "threshold control" block calculates the value of the accumulated MI and activates the appropriate detector, either the V-BLAST/ZF when the channel condition is good i.e. when the MI is above T_2; or the FSD during bad channel conditions, i.e. when MI is above T_1 but below T_2. Once the threshold is determined, the appropriate FPGA blocks are switched on and off accordingly. If the threshold falls under T_1, a re-transmission is required at a later time that consequently generates a new channel matrix, **H**, in the simulation process.

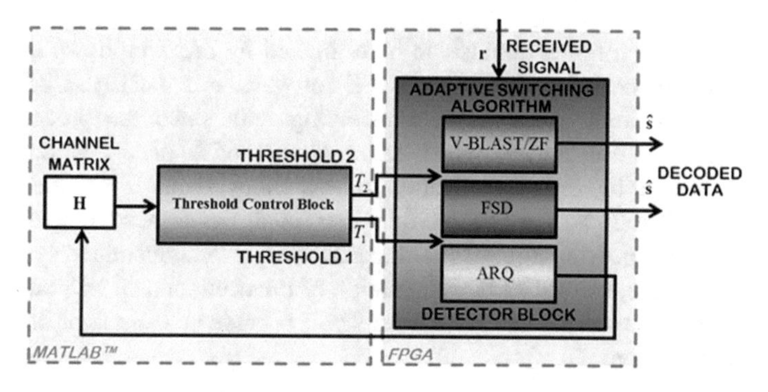

Fig. 2. Breakdown of Adaptive Switching Algorithm FPGA implementation model

For a complete receiver system, after the symbols are detected, they are passed to the turbo decoder for error correction. Although out of scope, the ongoing research shows that this complexity can be reduced by re-using the same MI calculation of the Adaptive Switching Algorithm in the detector to design the threshold for early termination of the turbo decoder. This will be detailed in the next paper of the author.

4 Results and Analysis

The energy performance analysis are based on the Xilinx® Virtex-7 chipset running at a core voltage of 1 V and an operating frequency of 250 MHz.

4.1 Part 1 - The Behavior of the Detector in Spatially Correlated Channels

The first part of the work involves in running separate detection algorithms that make up the Adaptive Switching Algorithm with different correlated channel factor. In order to investigate the impact they have on the channel correlation indexes, the channel correlations of **H** in Eq. (1) are set to be $\mathbf{R}_{Tx} = \mathbf{R}_{Rx} = \mathbf{C}$. The total resource allocation provided by the Xilinx® Integrated Synthesis Environment (ISE) for both detection algorithms is given in Table 1. The V-BLAST/ZF uses less resources, about a quarter of that required the more complex FSD.

Table 1. Xilinx® resource utilization for the V-BLAST/ZF and the FSD detection algorithms

XILINX® VIRTEX-7: XC7VLX330TFFG1157		
Logic resource utilization	Utilization	
	V-BLAST/ZF	FSD
Slice registers	3,312	13,683
Flip flops	892	4,688
4-input LUTs	2,940	12,161
DSP48E	48	132
Memory (RAM)	12	28

The number of multiplier counts can be estimated by breaking down the resource counter for each block using the Xilinx® ISE software. For V-BLAST/ZF, the most complexity comes from estimating the incoming data since the process requires complex matrix multiplications, which takes almost 65% of the whole detection algorithm, followed by data quantization of matching symbols on specific QAM constellation LUT at 26%. For FSD on the other hand, the highest complexity comes from calculating the metric, of the channel matrix against the transmitted symbols, uses most of the resources, as well as the summation of the accumulated ED, taking almost 75% of the total FSD operation. These results will provide an estimation for hardware design implementation.

When the two detection algorithms are implemented on different factors of **C**, the BER degrades significantly for both detection algorithms as depicted in Fig. 3(a) and (b) for FSD and V-BLAST/ZF respectively. As the channel correlation increases, getting more profound differences at higher SNR regions. This gets problematic at higher correlated channels when the V-BLAST/ZF is deployed, with BER of higher than 10^{-1} for $\mathbf{C} = 0.7$ for SNR ≤ 20 dB as depicted in Fig. 3(b). In order to achieve the BER tolerance design for the entire system of 10^{-3}, SNR approximately ≥ 45 dB for V-BLAST/ZF is required when the $\mathbf{C} = 0.7$ in comparison to SNR of approximately 27 dB for uncorrelated channels as depicted on Fig. 3(b). Similarly, a higher SNR is also needed or the FSD as shown in Fig. 3(a), where the BER for $\mathbf{C} = 0.7$, is also higher, at 10^{-2} for SNR of 20 dB and lower, and it requires an SNR of more than 26 dB to obey the system performance requirements. However, the BER performance would improve significantly when the turbo decoder is included in the design, which may help in dealing with maintaining the overall performance of the system on spatially correlated channels.

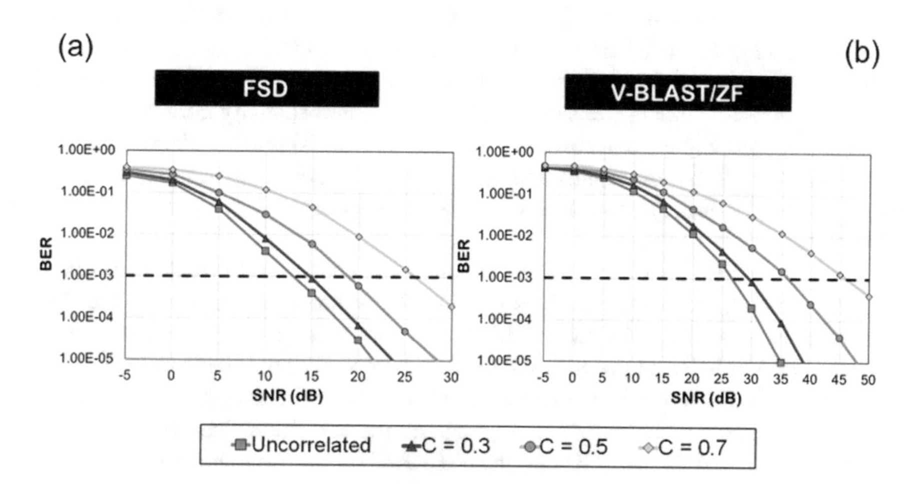

Fig. 3. Comparison of detector BER performance on spatially correlated channels for (a) FSD and (b) V-BLAST/ZF

With the performance verified, the MI values are calculated to provide the design of the thresholds for the Adaptive Switching Algorithm detector on different correlated channels. It is found that even though fading correlation does considerably affect the BER performance of each detection algorithm, the correlation index does not show any considerable changes to the MI values obtained.

Monte Carlo simulations are run 10 times, where each run comprises 100,000 channel realizations for each correlation index, **C**, at the SNR span of −5 dB to 20 dB. This can be observed in Fig. 4. More information on how the thresholds are determined can be found in [14].

The impact on the obtained MI thresholds shows only minor changes as the correlation of the channel increases. The two thresholds for the Adaptive Switching

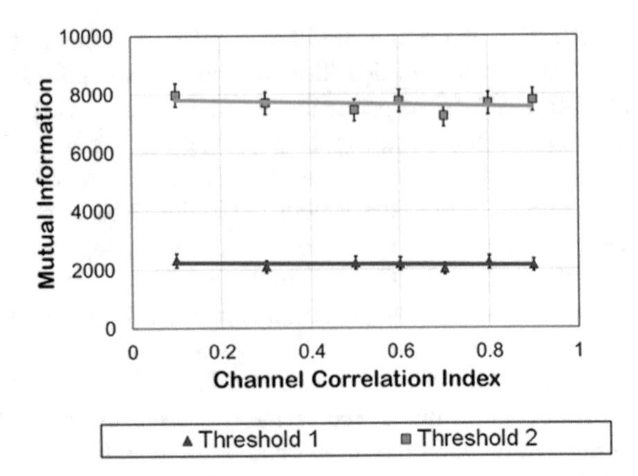

Fig. 4. Comparison of detector power consumption on spatially correlated channels

Algorithm detector lie in the range of 2,100 to 2,300 for, T_1, and 7,100 to 7,800 for threshold 2, T_2, for FSD and V-BLAST/ZF respectively. It gives a linear trend therefore, it can be concluded that the threshold values for the Adaptive Switching Algorithm detector remain the same even when applied spatially correlated channels and it can be said that the detector design is only specific to the modulation and coding schemes in use. With these results, the design for the proposed algorithm is set as 2,200 and 7,100 for T_1 and T_2 respectively. T_1 corresponds to the BER = 0.5 and T_2 for a BER of 10^{-3}.

The other performance parameter, which is the energy consumption, can be calculated by taking the power readings provided by Xilinx® ISE and using the time it takes to transfer a packet bit size of 1,024 at a core voltage of 1 V and an operating frequency of 250 MHz on the Xilinx® Virtex-7 chipset. For the span of the SNR levels of −5 dB to 20 dB, the average energy consumption of the two detection algorithms within the Adaptive Switching Algorithm against the correlated channel index range of 0 to 1 are computed for the FSD and the V-BLAST/ZF as 3.6 J and 0.9 J respectively. This shows that with the increase in correlation, the energy consumption of the detector is hardly affected as well. This could be due to the both algorithms work independently of noise level and have a fixed distinct search on any channel conditions. For the detector, it can be concluded that, comparable energy savings can be gained in spatially correlated channels as well. When combining both algorithms to make the Adaptive Switching Algorithm, Fig. 5 shows the energy consumption on a spatially correlated channels. In the detector, the energy savings when utilizing the Adaptive Switching Algorithm on different correlated channel indexes can be calculated numerically for SNR. range of 0 dB to 50 dB for a run of 100,000 channel realizations on the chosen hardware. This is essentially the area under the graph of Fig. 5 if the FSD is taken as the 100% baseline at 3.6 J. The results are tabulated in Table 2. It can be observed that though there are still savings gained, the energy savings decreases with higher channel correlation.

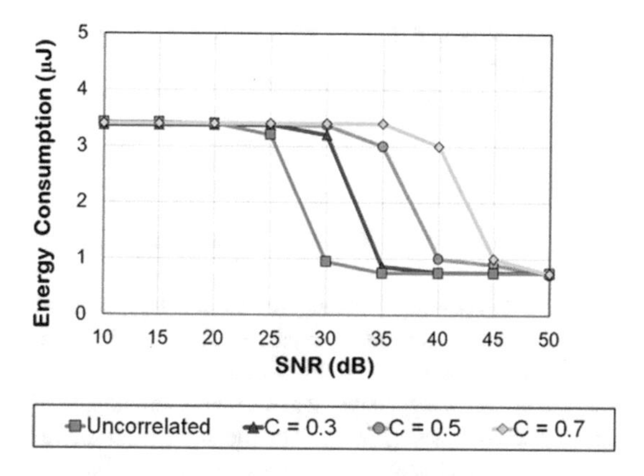

Fig. 5. Energy consumption of the adaptive switching algorithm detector in spatially correlated channels

Table 2. Energy savings of Adaptive Switching Algorithm detector on spatially correlated channels

Correlation index (C)	Energy savings (%)
Uncorrelated	40%
0.3	33%
0.5	27%
0.7	19%

Figure 5 also shows the reason for the reduced energy saving, which is that, the threshold T_2 between the two algorithms corresponds to a much higher SNR for higher channel correlation values. From the figure, it can be observed that the switching occur an SNR 25 dB for uncorrelated channels, and SNR 46 dB for $\mathbf{C} = 0.7$. It can be concluded that, the energy usage varies for the Adaptive Switching Algorithm with varying channel correlation factors, with lower savings can be gained as the correlation increases.

5 Conclusion

The Adaptive Switching Algorithm was utilized in the first part of the MIMO receiver. The detector uses the threshold calculations involving the MI between the transmitters and receivers provide sufficient information in real-time regarding any channel conditions, whether uncorrelated or spatially correlated. The work has proven that the average energy savings in the detector can be achieved throughout the span of considered SNR conditions of −5 dB to 20 dB, and they are to be at the range of 19% to 40% when implemented on Xilinx® Virtex-7 chipset. Further work has been implemented to show that the design for the Adaptive Switching Algorithm can be expanded to be a link between the detector and decoder, and proved to save even more energy. For more detailed description, refer to author's future work.

Acknowledgment. The author would like to send her deepest gratitude to University Tun Hussein Onn Malaysia for sponsoring the research.

References

1. Telatar, E.: Capacity of multi-antenna Gaussian channels. Eur. Trans. Telecommun. **10**(6), 585–595 (1999)
2. Foschini, G.J.: Layered space-time architecture for wireless communication in a fading environment when using multi-element antennas. J. Bell Lab. Technol. **1**(2), 41–59 (2002)
3. Shiu, D.S., Foschini, G.J., Gans, M.J., Kahn, M.: Fading correlation and its effect on the capacity of multielement antenna systems. IEEE Trans. Commun. **48**(3), 502–513 (2000)
4. Wubber, D., Kulm, V., Kammeyer, K.D.: On the robustness of lattice-reduction aided detectors in correlated MIMO. In: Proceedings of IEEE 60th Vehicular Technology Conference, vol. 5, no. 1, pp. 3639–3643, September 2004
5. Meng, Q., Pan, Z., You, X., Kim, Y.H.: On performance of lattice reduction aided detection in the presence of receive correlation. In: Proceedings of IEEE 6th Circuits and Systems Symposium on Emerging Technologies: Frontiers of Mobile and Wireless Communications, Transactions on Information Theory, vol. I, no. 1, pp. 89–92, June 2004
6. Barbero, L.G., Thompson, J.S.: Performance of the complex sphere decoder in spatially correlated channel. J. Inst. Eng. Technol. **1**(1), 122–130 (2007)
7. Proakis, J.G.: Digital Communications, 3rd edn., pp. 767–768. McGraw-Hill Book Co., Singapore (1983)
8. Sklar, B.: Rayleigh fading channels in mobile digital communication systems Part I: characterization. IEEE Commun. Mag. **35**(7), 90–100 (1997)
9. Goldsmith, A., Biglieri, E., Calderbank, R.: MIMO Wireless Communications, p. 559. Stanford University, Stanford (2007)
10. Kermoal, J., Schumacher, L., Pedersen, K.I., Mogensen, P., Frederiksen, F.: A stochastic MIMO radio channel model with experimental validation. IEEE J. Sel. Areas Commun. **20**, 1211–1226 (2002)
11. Yu, K., Bengtsson, M., Ottersten, B., McNamara, D., Karlsson, P., Beach, M.: Modeling of wide-band MIMO radio channels based on NLoS indoor measurements. IEEE Trans. Veh. Technol. **53**, 655–665 (2004)
12. Tulino, A.M., Lozano, A., Verdu, S.: Impact of antenna correlation on the capacity of multiantenna channels. IEEE Trans. Inf. Theory **51**(7), 2491–2509 (2005)
13. Ozcelik, H., Herdin, M., Weichselberger, W., Wallace, J., Bonek, E.: Deficiencies of 'Kronecker' MIMO radio channel model. IEEE Electron. Lett. **36**(16), 1209–1210 (2003)
14. Tadza, N., Laurenson, D.I., Thompson, J.S.: Adaptive Switching Detection algorithm for iterative-MIMO systems to enable power savings. J. Radio Sci. **49**(11), 1065–1079 (2014)
15. Barbero, L.G., Thompson, J.S.: Fixing the complexity of the sphere decoder for MIMO detection. IEEE Trans. Wireless Commun. **7**(6), 2131–2142 (2008)
16. Golden, G.D., Foschini, C.J., Valenzuela, R.A., Wolniansky, P.W.: Detection algorithm and initial laboratory results using V-BLAST space time communication architecture. IEEE Electron. Lett. **35**(1), 14–16 (1999)
17. Tadza, N., Laurenson, D.I., Thompson, J.S.: Power performance analysis of the iterative-MIMO adaptive switching algorithm detector on the FPGA hardware. In: IEEE Vehicular Technology Conference (VTC), May 2015. http://tinyurl.com/on6wqle

Implementation of Precision Deburring System Using Dissolved Gas Control

Young-Dong Lee[1] and Do-Un Jeong[2(✉)]

[1] Department of Computer Software Engineering, Changshin University, Changwon, Republic of Korea
[2] Department of Ubiquitous IT Engineering Graduate School, Dongseo University, Busan, Republic of Korea
dujeong@dongseo.ac.kr

Abstract. In this study, we implemented an ultrasonic deburring system to remove the burr formed after the casting process of the valve body which is a key part in the manufacturing process of automobile and machine parts. The ultrasonic deburring system consists of a vibrating part and a control part. To maximize the deburring effect, the dissolved gas control method included in the deburring water is applied. Experimental results show that different types of burrs are removed, and it is confirmed that this method can be applied to real industry.

Keywords: Deburring · Valve body · Ultrasonic deburring · Dissolved gas

1 Introduction

Burr means foreign substances which are produced in cutting or processing of a material, and it is accompanied by a process for removing the burrs in the manufacturing process. A deburring and cleaning process is performed to remove burrs from the manufacturing process of the valve body which is a core part of the engine and the transmission of the automobile. The conventional burr removal method uses a vibration method using a steel ball of 2 pie. In this case, it is difficult to precisely remove the minute burrs of the valve body of a complicated structure by using only the left and right vibrations. In addition, there is a concern that not only the deburring takes a long time but also the defect rate increases and the life of the equipment is shortened due to the insertion of the steel ball into the corner groove. If the burrs are not completely removed and the missions are assembled in the grooved grooves, it can lead to fatal defects such as pathway clogging of the mission oil, or valve failure.

In this study, ultrasonic deburring system was developed to remove burrs more efficiently and to simplify the cleaning process. In order to improve the deburring performance, a degassing device for controlling the dissolved gas in the deburring water was constructed and interlocked with the deburring device to maximize its performance. For the evaluation of the performance, the burrs were artificially created on the metal, and the deburring results were observed with a microscope before and after the removal.

K.J. Kim et al. (eds.), *IT Convergence and Security 2017*,
Lecture Notes in Electrical Engineering 450,
DOI 10.1007/978-981-10-6454-8_17

2 Cavitation

In this study, we propose a system capable of precise deburring using ultrasonic. An important factor for precise deburring is the density of cavitation generated using ultrasonic generators. It is necessary to generate cavitation of a high density through variable frequency using an ultrasonic oscillator in a variable frequency system or a plurality of generators. The strength and density characteristics of the cavity according to the shape and frequency of the cavity according to the oscillation frequency are shown in Figs. 1 and 2 respectively.

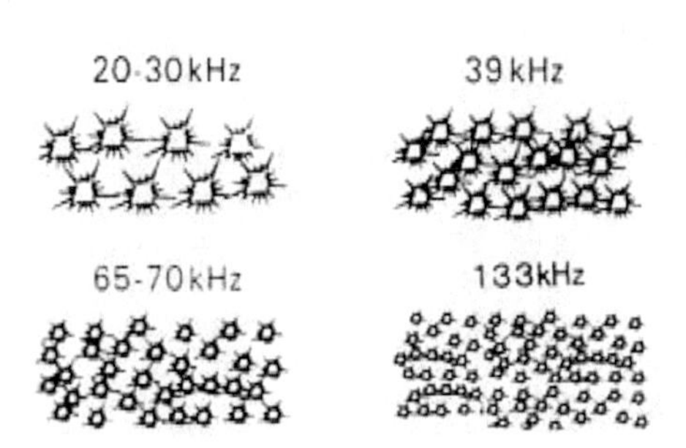

Fig. 1. Cavity shape by frequency.

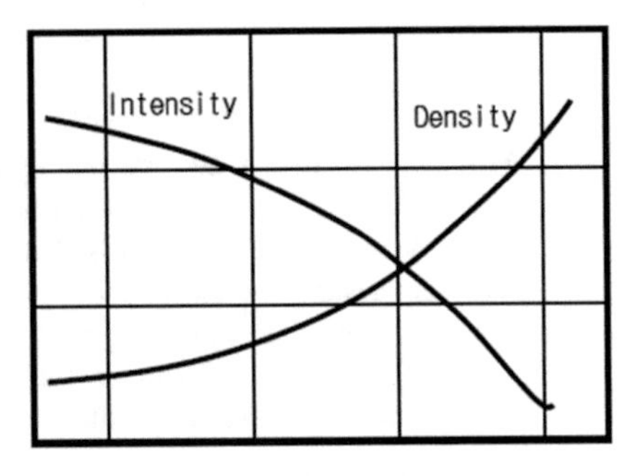

Fig. 2. The strength and density of cavities according to frequency.

3 Ultrasonic Vibrator

In this study, ultrasonic oscillator is placed on the side and bottom to allow variable selection of ultrasonic generation position, and ultrasonic oscillator of 20 and 28 kHz band which can be deburred and cleaned is applied. Figures 3 and 4 show the results of arranging the oscillator and the oscillator in this study.

Fig. 3. Ultrasonic vibrator.

Fig. 4. An oscillator arranged in an array structure.

The configuration of the ultrasonic oscillator for driving the ultrasonic transducer enables the generation of the variable frequency through one oscillating system and the system is implemented so that a plurality of oscillators can be constructed using one oscillation module.

4 Experimental Result

The ultrasonic transducer implemented in this study was attached to the deburring water tank. To maximize the efficiency of deburring, the transducer modules were arranged in the side and bottom of the water tank. Deaerator was applied to generate deaeration water to improve the performance of the deburring and the deburring performance was evaluated by the concentration of the dissolved gas. Figure 5 shows an ultrasonic vibrator mounted on a deburring water tank implemented in this study. Figure 6 shows a variable frequency oscillator for driving an ultrasonic vibrator.

Fig. 5. Deburring tank.

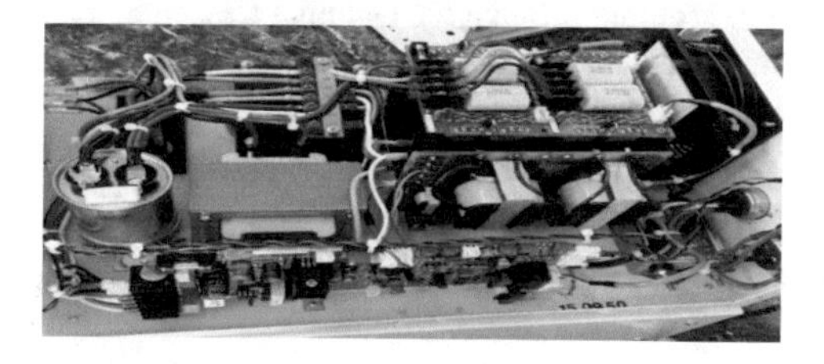

Fig. 6. Variable frequency oscillation system.

In order to maximize the efficiency of ultrasonic deburring, dissolved gas control of deburring water is important and a degassing system is applied for this purpose.

Additionally, an artificial burr was formed on a number of metals to improve the deburring efficiency by controlling the dissolved gas. Figure 7 shows the performance evaluation test of the ultrasonic deburring system and dissolved gas control.

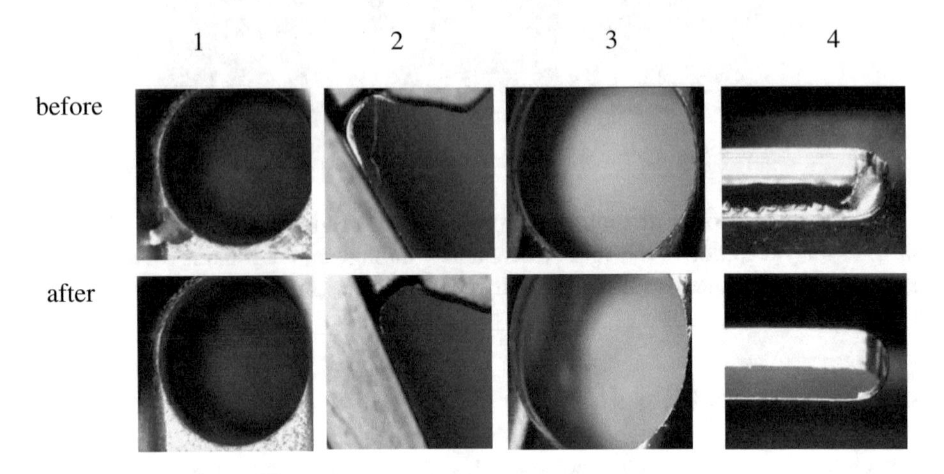

Fig. 7. Deburring performance evaluation.

5 Conclusion

In this paper, we developed a precise ultrasonic deburring system with dissolved gas control and evaluated the performance of the developed deburring system. As a result of experiments, it was verified through experiments that more efficient deburring is possible through control of dissolved gas. We will carry out further experiments to objectively evaluate the deburring performance of various conditions and the degree of elevation of deburring efficiency according to dissolved gas concentration.

Acknowledgements. This research was supported by a research program of Dongseo University's Ubiquitous Appliance Regional Innovation Center supported by the grants from Ministry of Commerce, Industry and Energy (No. B0008352). And This research was supported by Basic Science Research Program through the National Research Foundation of Korea (NRF) funded by the Ministry of Education (No. 2015R1D1A1A01061131).

References

1. Yanru, H., et al.: Study of deburring machine automatic pineapple peel and color sensor. JAMR **3**, 037 (2015)
2. Valiev, R.I., et al.: Polishing and deburring of machine parts in plasma of glow discharge. In: IOP Conference Series, vol. 86, no. 1. IOP Publishing (2015)
3. Mori, M., Nagasuna, T., Hamada, H.: The difference in micro-deburring finish produced by groove cutting method. In: International Conference on Digital Human Modeling and Applications in Health, Safety, Ergonomics and Risk Management. Springer (2016)

Cache Reuse Aware Replacement Policy for Improving GPU Cache Performance

Dong Oh Son[1], Gwang Bok Kim[2], Jong Myon Kim[3], and Cheol Hong Kim[2(✉)]

[1] Doosan FuelCell Power BU, 75 Jeyakdanji-Ro, Hwaseong-si, Korea
Sdo1127@gmail.com

[2] Chonnam National University, 77 Yongbong-Ro, Buk-Gu, Gwangju, Korea
loopaz63@gmail.com, chkim22@jnu.ac.kr

[3] University of Ulsan, 93 Daehak-Ro, Nam-Gu, Ulsan, Korea
jongmyon.kim@gmail.com

Abstract. The performance of computing systems has been improved significantly for several decades. However, increasing the throughput of recent CPUs (Central Processing Units) is restricted by power consumption and thermal issues. GPUs (Graphics Processing Units) are recognized as efficient computing platform with powerful hardware resources to support CPUs in computing systems. Unlike CPUs, there is a large number of CUDA (Compute Unified Device Architecture) cores in GPUs, hence, some cache blocks are referenced many times repeatedly. If those cache blocks reside in the cache for long time, hit rates can be improved. On the other hand, many cache blocks are referenced only once and never referenced again in the cache. These blocks waste cache memory space, resulting in reduced GPU performance. Conventional LRU replacement policy cannot consider the problems from non-reused cache blocks and frequently-reused cache blocks. In this paper, a new cache replacement policy based on the reuse pattern of cache blocks is proposed. The proposed cache replacement policy manages cache blocks by separating reused cache blocks and thrashing cache blocks. According to simulation results, the proposed cache reuse replacement policy can increase IPC by up to 4.4% compared to the conventional GPU architecture.

Keywords: GPGPU · Cache · Cache reuse · Replacement policy · Performance

1 Introduction

Nowadays, the performance of computing systems has been constrained. The main reason behind is that techniques for increasing the throughput of CPUs (Central Processing Units) are restricted by power consumption and thermal issues. Meanwhile, GPUs are recognized as a powerful computing platform with large hardware resources because of their ability in processing a plurality of threads in parallel [1]. Therefore, one of solutions to this problem is to share the workload of CPUs with GPUs (Graphics Processing Units). As a result, the performance of computing systems can be improved [2–4].

K.J. Kim et al. (eds.), *IT Convergence and Security 2017*,
Lecture Notes in Electrical Engineering 450,
DOI 10.1007/978-981-10-6454-8_18

Cache memory is used as a buffer to hide the speed gap between processor and main memory. Such kind of memory is usually organized in associativity (set-associative mapping or fully-associative mapping) instead of direct mapping. GPU cache is divided into L1 and L2 cache. L1 cache is used to store data referenced by CUDA cores. Meanwhile, L2 cache stores data referenced by SMs. On GPUs, there are many types of L1 caches such as instruction, data, texture and constant. Recently, for improving performance of GPGPUs, scheduling methods, internal networks and data parallelism enhancement techniques have been studied [5–7]. However, GPU cache has not received enough attention. Previous research works showed that improving the performance of cache architecture can lead to high performance of GPGPUs [8, 9]. Therefore, in this paper, we concentrate on GPU cache architecture to improve the GPU performance. We propose a cache replacement policy that is based on the characteristics of GPUs. The cache replacement policy used in GPU cache is LRU, which is also commonly-used in CPUs. However, on GPU cache architectures, there some cache blocks are referenced many times due to the large number of CUDA cores. Cache replacement policy is designed based on temporal locality (blocks are referenced in a small time duration) and spatial locality (the use of blocks in close locations). Therefore, cache replacement policy is important to cache architecture, especially when cache size is small. In fact, previous studies showed that LRU replacement policy cannot provide good performance for memory intensive workloads which usually have a large working set than the cache size. When working set is larger than cache size, cache blocks will be replaced before they are referenced again, known as thrashing issue. Other works like DIP (Dynamic Insertion Policy) and RRIP (Re-Reference Interval Prediction) can solve the thrashing issue in CPUs [10, 11].

The rest of this paper is organized as follows. Section 2 explains more about the proposed cache replacement policy and existing techniques. Section 3 describes in detail experimental environment and results, and finally Sect. 4 concludes this paper.

2 Cache Reuse Aware (CRA) Replacement Policy

2.1 Cache in GPU Architecture

Although GPU cache architecture is very similar to CPU cache, the hit rate is different because GPU cache does not consider architectural characteristics. CPU cache tries to predict blocks that may be reused by using spatial and temporal locality. However, as GPU has many CUDA cores, there may be many requests to the same cache block at different times. If a CUDA core requests a cache block, the probability that other CUDA cores also request the cache block is high because CUDA cores in an SM execute the same instruction at one time. However, GPU cache architecture does not care this characteristic, resulting in low cache hit rate.

Figure 1 shows cache reuse cycle with ‘Back Propagation’ application [12]. There is 57.9% of reused cache blocks having the reused period of 1 to 2 cycles and 42.1% having the reused period is greater than 2 (long). Conventional LRU replacement policy always replaces blocks that have long reuse period. In this paper, we consider not to always replace such cache blocks. The L1 data cache is modeled as an 8-way set associative, write-through, no write-allocate cache. Only blocks that have at least once reused can apply this proposed method while blocks have more than 8 times of reuse cannot. We set 2 ways for thrashing part (6 ways for reuse part) because the cache reuse cycle probability from 1 to 2 is high.

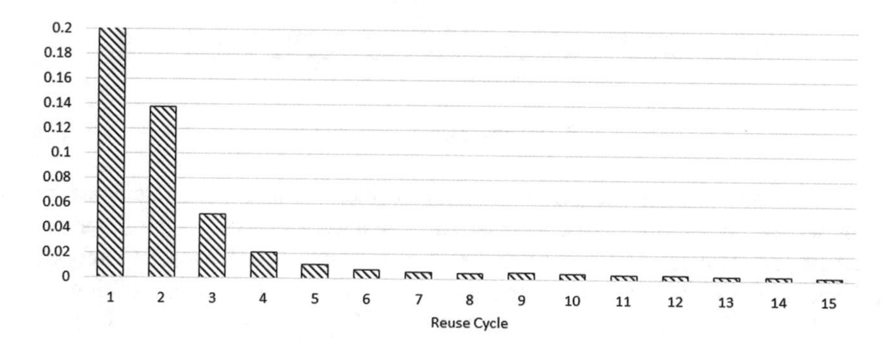

Fig. 1. Performance according to L1 data cache size

2.2 Proposed Cache Replacement Policy

In this paper, we propose a new cache replacement policy that considers the GPU characteristics, called Cache Reuse Aware (CRA) replacement policy. CRA replacement policy assigns reused cache blocks to reuse part for longer residing in cache. This is because these reused cache block has high reuse probability in the future. Thus, CRA replacement policy manages reused blocks to improve performance.

In order to improve the cache hit rate have been proposed various cache replacement policy. There are many commonly-used cache replacement policies for improving cache hit rate including Random, FIFO, LRU, LFU, and NRU (Not Recently Used). Some recently proposed replacement policies are recognized good performance such as DIP, RRIP. However, LRU is the most popular policy used in CPU and GPU. If a CUDA core requests a cache block, the probability that other CUDA cores also request the cache block is high since CUDA cores in an SM execute the same instruction at one time. However, many cache blocks are referenced only one time and eventually replaced, this can degrade performance. On CPU, each cache block has the same opportunity and the least recently used cache block is evicted. This method is called LRU replacement policy, however, applying LRU in GPU cache is not effective.

The Fig. 2 shows the comparison of CRA replacement policy to LRU replacement policy. Diagonal stripes blocks refer the blocks that are non-reused and the non-diagonal stripes block refers to reused blocks. Conventional LRU replacement policy does not consider non-reused and reused blocks. Hence, if a reused block is evicted and requested again, this cause a cache miss. Separating cache blocks into two groups, the left side stores reused cache blocks and the right side applies LRU for thrashing cache blocks may be the solution to the problem. This method can replace non-reused cache blocks and the reused blocks can reside in cache for a long time.

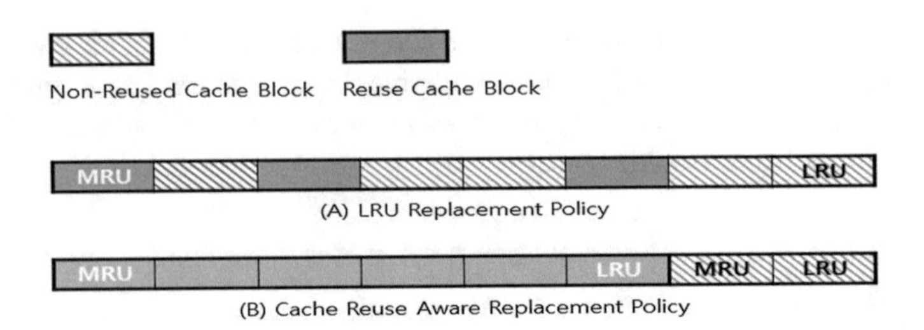

Fig. 2. Comparison of CRA to LRU

Figure 3 shows the principle of CRA replacement policy. The reused part which stores reused cache blocks also applies LRU replacement. If a cache block in the thrashing part is hit, it will be moved to the reuse part at the MRU (Most Recently Used) position (1) the LRU block of the reuse part will be moved to MRU position of the thrashing part (2) thrashing part has LRU and MRU positions. A new cache block is inserted into MRU position. On a miss, the LRU block of the thrashing part is replaced.

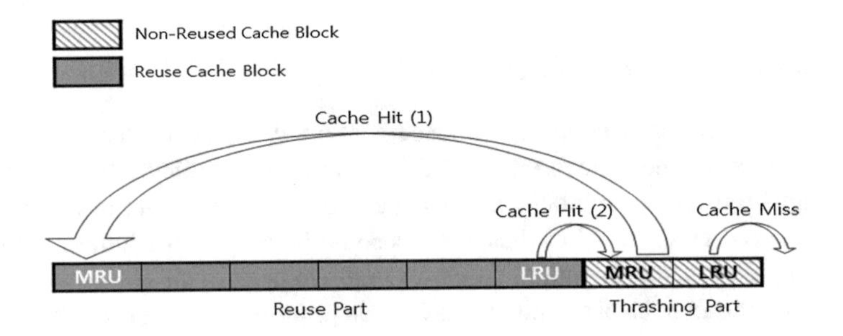

Fig. 3. Cache reuse aware replacement policy

3 Evaluation

3.1 Experimental Methodology

In this section, we briefly describe our experimental methodology. To analyze the GPU performance according to the proposed CRA replacement policy, various parallel applications are selected from NVIDIA SDK [13], ISPASS [14]. In this paper, modified GPUWattch constructed by GPGPU-Sim and McPAT to evaluate the performance and energy savings of proposed techniques. To execute selected benchmarks, we use the GPGPU-sim to model diverse aspects of GPU. Table 1 presents the system configuration of baseline GPU used in our experiments.

Table 1. System configuration

Parameter	Value
Number of SM	15
Warp size (SIMD width)	32
Number of threads/SM	1024
Shared memory/SM	16 KB
Constant cache/SM	8 KB, 2-way 64 byte lines, Read-only
Texture cache/SM	12 KB, 24-way 128 byte lines, Read-only
L1 data cache	32 KB, 8-way, 128 byte lines, Read-only
Unified L2 cache	64 KB, 8-way, 128 byte lines
Clock (core: interconnection: DRAM)	575 MHz: 575 MHz: 750 MHz
Number of memory controller	6
Number of memory chip/controller	2
CTA&warp scheduler	Two-level scheduler (Round-Robin)

3.2 Experimental Results

In this section, we analyze the GPU performance when multiple-applications are executed in parallel. The applications are selected from NVIDIA SDK [13] including Discrete Cosine Transform (DCT), and ISPASS [14] including Breadth-First Search (BFS), LIBOR Monte Carlo (LIB), Ray Tracing (RAY), Weather Prediction (WP).

Compared to LRU replacement policy, CRA replacement policy is able to keep the reuse cache blocks, and thus reduces victim cache blocks. Figure 4 shows the GPU performance of the proposed CRA replacement policy for various applications. This figure presents normalized performance (IPC) improvement of CRA replacement policy normalized to the baseline replacement policy. CRA replacement policy shows higher IPC compared to LRU replacement policy. The performance of GPGPU is increased considerably, 2.8% on average (4.4% for BFS, 2.3% for DCT, 1.3% for LIB, 3.6% for RAY, 2.4% for WP). This means that CRA replacement policy can better increase cache hit rate than baseline replacement policy. This is because CRA replacement policy optimizes cache area assigned according to thrashing issue.

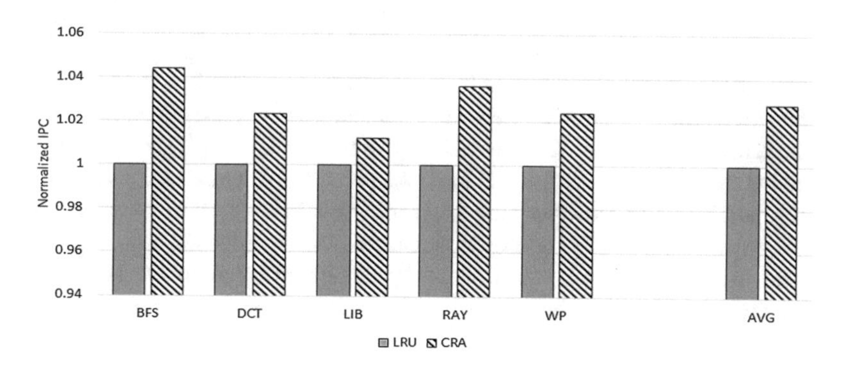

Fig. 4. Performance of CRA replacement policy (IPC)

To analyze the cache performance according to CRA replacement policy, we evaluate L1 data cache hit rate. Figure 5 shows that CRA replacement policy significantly improves cache hit rates for various applications.

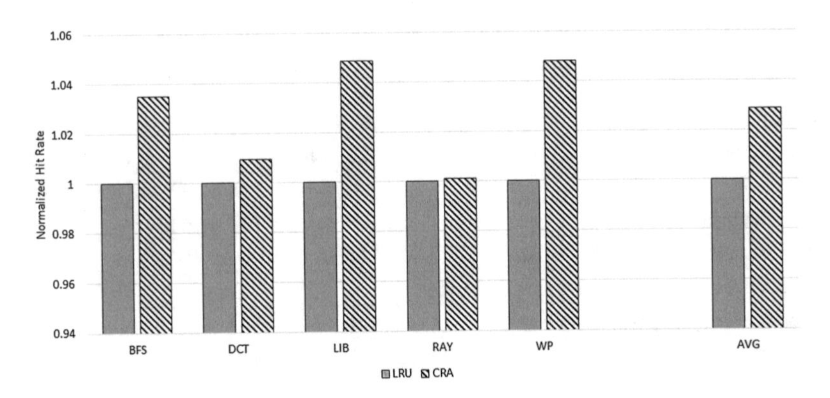

Fig. 5. Cache performance of CRA replacement policy

The reason for the hit rates improvement for CRA replacement policy is an increase in cache hit rates. For example, more than a 4.9% increase is observed over CRA replacement policy for LIB application compared to LRU. The miss rate of various applications declines for CRA replacement policy because reuse part can effectively store reuse cache blocks. Thus, the reuse cache blocks are stored from eviction when thrashing. CRA replacement policy can achieve better GPU L1 data cache hit rate over LRU. On average, the hit rate improvement is about 2.9% (3.5% for BFS, 1.0% for DCT, 4.9% for LIB, 0.1% for RAY, 4.8% for WP). From these results, we find that CRA replacement policy can increase cache hit rate better than LRU replacement policy.

4 Conclusions

In this paper, a new cache replacement policy for the GPGPU architecture is proposed. Note that the LRU policy is used as the baseline cache replacement policy. Comparing to the traditional cache replacement policy, the proposed cache replacement policy provides better performance for the GPU. The conventional LRU policy is based on the assumption that a cache block would be re-accessed soon after it was accessed in the CPU. However, the conventional LRU replacement policy is not effective in the GPU since the reuse pattern in the GPU is different to that in the CPU. The proposed CRA replacement policy divides cache sets into two parts, one is reuse part and the other is thrashing part. If a cache block is reused, it is moved to the reuse part, hence solving the thrashing issue. According to the experimental results, the proposed cache replacement policy can improve average performance by 4.4% compared to the baseline LRU replacement policy.

Acknowledgement. This research was supported by the MSIP (Ministry of Science, ICT and Future Planning), Korea, under the ITRC (Information Technology Research Center) support program (IITP-2016-R2718-16-0011) supervised by the IITP (Institute for Information & communications Technology Promotion)

References

1. Buck, I., Foley, T., Horn, D., Sugerman, J., Fatahalian, K., Houston, M., Hanrahan, P.: Brook for GPUs: stream computing on graphics hardware. In: Proceedings of 31st Annual Conference on Computer Graphics, Los Angeles, pp. 777–786 (2004)
2. General-purpose computation on graphics hardware. http://www.gpgpu.org/
3. Owens, J.D., Luebke, D., Govindaraju, N., Harris, M., Kruger, J., Lefohn, A.E., Purcell, T.J: A survey of general-purpose computation on graphics hardware. In: Euro-graphics, pp. 21–25 (2005)
4. Yang, Y., Xiang, P., Mantor, M., Zhou, H.: CPU-assisted GPGPU on fused CPU-GPU architectures. In: High-Performance Computer Architecture, pp. 1–12 (2012)
5. Lee, S.Y., Arunkumar, A., Wu, C.J.: Coordinated warp scheduling and cache prioritization for critical warp acceleration of GPGPU workloads. In: Proceedings of International Symposium on Computer Architecture, pp. 515–527 (2015)
6. Kayiran, O., Nachiappan, N.C., Jog, A., Ausavarungnirun, R., Kandemir, M.T., Loh, G.H., Das, C.R.: Managing GPU Concurrency in Heterogeneous Architectures. In: Proceedings of International Symposium on Microarchitecture, Cambridge, vol. 43, pp. 114–126 (2015)
7. Wang, J., Rubin, N., Sidelnik, A., Yalamanchili, S.: Dynamic thread block launch: a lightweight execution mechanism to support irregular applications on GPUs. In: Proceedings of International Symposium on Computer Architecture, pp. 528–540 (2015)
8. Liang, Y., Wang, Y., Sun, G.: Coordinated static and dynamic cache bypassing for GPUs. In: Proceedings of the International Symposium on High-Performance Computer Architecture, pp. 76–88 (2015)
9. Singh, I., Shriraman, A., Fung, W.W., O'Connor, M., Aamodt, T.M: Cache coherence for GPU architectures. In: Proceedings of the International Symposium on High-Performance Computer Architecture, pp. 578–590 (2013)
10. Qureshi, M.K., Jaleel, A., Patt, Y.N., Steely, S.C., Emer, J.: Adaptive insertion policies for high performance caching. In: Proceedings of International Symposium on Computer architecture, vol. 35, pp. 381–391 (2007)
11. Jaleel, A., Theobald, K.B., Steely Jr., S.C., Emer, J.: High performance cache replacement using Re-Reference Interval Prediction (RRIP). In: Proceedings of International Symposium on Computer Architecture, pp. 60–71 (2010)
12. Goodrum, M.A., Trotter, M.J., Aksel, A., Acton, S.T., Skadron, K.: Parallelization of particle filter algorithms. In: Proceedings of the International Symposium Computer Architecture, pp. 139–149. Springer, Heidelberg (2010)
13. NVIDIA CUDA SDK. http://developer.download.nvidia.com/compute/cuda/sdk/
14. Bakhoda, A., Yuan, G.L., Fung, W.W., Wong, H., Aamodt, T.M.: Analyzing CUDA workloads using a detailed GPU simulator. In: Proceedings of International Symposium on Performance Analysis of Systems and Software, pp. 163–174. IEEE Press, Boston (2009)

Security and Privacy

Security Perspective in Comparative Study of Platform-Based and Platform-Less BYOD Systems

Nithin R. Aenugu, Sergey Butakov(✉), Pavol Zavarsky, and Shaun Aghili

Information Systems Security and Assurance Management, Concordia University of Edmonton, Edmonton, Canada
naenugu@student.concordia.ab.ca, {sergey.butakov, pavol.zavarsky, shaun.aghili}@concordia.ab.ca

Abstract. Two different ways of adopting Bring Your Own Device (BYOD) policy in an enterprise are either contracting it out to vendor or building custom made BYOD solution with a set of independent security mechanisms and organizational policies. A comparative analysis of the two types of BYOD solutions has been performed to assist enterprises that consider ways in adoption of BYOD policy other than contracting it out to a vendor. In addition to the comparative analysis, this research outlines residual risks to help enterprises determine the suitable type of BYOD solution for their needs. To analyze BYOD risks, major risk areas such as lost & stolen devices, unauthorized third party applications, etc. have been analyzed. For all these risk areas, risk mitigation techniques were outlined and residual risks have been analyzed using risk scenarios derived from widely accepted industry standards suggesting the best practices. This research work adds as an additional source document to the knowledgebase of BYOD.

1 Introduction

Adoption of BYOD policy in enterprises sees a considerable rise in the last decade and the same trend is expected in the nearest future [3]. Gartner in 2013 estimated that 38% of large corporations are willing to allow employees to bring their own devices for work purposes solely or as a reinforcement [9]. If the adoption of BYOD is not properly managed by an enterprise in line with enterprise infosec policies, such an adoption can lead to potential loss of reputation, misuse of enterprise assets and financial loss [1]. The choice to select the type of BYOD solution depends on the enterprise's current risk profile. Size and sector of the enterprise helps evaluate the requirements depending on the sensitivity of the data and risk by exposure of resources based on the environment they operate in. This information is further useful in arriving at decision of choosing the type of BYOD solution [17]. Other factors that affect the decision are geographic deployment, company's compliance requirements, as well as training and support [10].

K.J. Kim et al. (eds.), *IT Convergence and Security 2017*,
Lecture Notes in Electrical Engineering 450,
DOI 10.1007/978-981-10-6454-8_19

While adoption of BYOD in enterprises is gradually increasing with proliferation of mobile devices at work place, enterprises however are only limited with a limited range of options [7]. In this research, vendor based BYOD solutions are referred as Platform-based BYOD solution, and assembled set of independent security mechanisms is called Platform-less BYOD solution. BYOD solutions available in the market today are vendor provided and they received noticeable attention from the industry experts [7]. Less attention paid on platform-less BYOD solutions is identified as one of the reasons that is limiting the range of BYOD solutions available. Although platform-less BYOD solutions are the kind that the industry has first seen, platform-based BYOD solutions overtook them and began to lead BYOD market with the rise of Mobile Device Management (MDM) solutions [10].

To present all the possible options of deployment of BYOD, a closer look at the platform-less BYOD solution helps change the way enterprises look at their options when considering adoption of BYOD policy. It is obvious that platform-less BYOD solution will require significant level of expertise from the enterprise. However, detailed side – by – side comparison of platform-based and platform-less BYOD solutions will help to determine whether or not it is feasible to develop a platform-less BYOD solution within an enterprise without compromising objectives of BYOD.

This research provides a comparative analysis of platform-based and platform-less BYOD solutions aiming at presenting how these solutions handle different kind of risks that may arise in BYOD environment. Also, residual risks that may remain after implementation of risk mitigation techniques solely or as combination of two or more techniques are studied to identify whether it is worth adopting platform-based or platform-less BYOD solution.

2 Related Works

There are few theoretical frameworks that recommend strategies to mitigate risks associated with BYOD. The review covers BYOD strategies that have been outlined in peer reviewed research literature as well as in some standards and good practices released by Information Systems Audit and Control Association (ISACA) and National Institute of Standards and Technology (NIST). Analysis of peer-reviewed literature has also helped understand how policy makers can enforce some security measures to control certain risks. To further proceed with, platform-based BYOD solutions such as Samsung KNOX and MobileIron have been studied to identify the strategies that are being adopted by providers at industry level. Study of such strategies is a useful entity to find out whether it is worthy for an enterprise to invest resources in adopting a Platform-less BYOD solution, if they follow this route.

Srikant et al. suggests that the concept of containerization that KNOX is equipped with should be one of the key concepts that enterprises have to consider in developing a BYOD solution for an enterprise [1]. The author has mentioned that containerization is also effective in the event of remote wipe which is capable of wiping off just the enterprises' information without touching employee's personal data [1]. This is the key strategy that allows BYOD solutions to comply the policies protecting employees' privacy while satisfying some needs to protect enterprise data. Similar to KNOX, most

MDM solutions have adopted containerization, also called *sandboxing*. This functionality basically isolates business data holding it separate from personal data on the device to create a BYOD friendly environment.

MDM solutions in work environment serve most of the BYOD objectives. Since MDM solutions have seen significant rise in development due to proliferation of mobile devices at work places, enterprises are forced to depend on the solutions for BYOD purposes. Various IT giants have come up with MDM solutions such as Samsung KNOX. On the other hand, research work on non-enterprise specific solutions has led to recommendations and frameworks for enterprises in developing BYOD solutions. Various industry recommendations suggest that security mechanisms such as encryption, two-factor authentication and remote wipe systems should be considered as key tools in choosing BYOD platform [6].

A case study by SANS [5] talks about MDM agent to control the access to the device and the wireless network. It mentioned that with the help of strong operational policies a list of exempted devices that are excluded from access control policy needs to be maintained [5]. Author has also mentioned that a combination of access control policy and MDM system can create a full BYOD platform with acceptable risk [5].

A security perspective look at both type of BYOD solutions will help an enterprise to evaluate the best suitable type of solution on their own. Additionally, by evaluating BYOD requirements with respect to the enterprise risk profile will also help having a knowledge of best practices to follow in a BYOD environment.

2.1 Existing BYOD Frameworks

To establish uniform management framework and to provide guidance on how to embed security for mobile devices into IT governance, risk and compliance (GRC), ISACA have published a framework called *Securing Mobile Devices*. Related document - *COBIT 5 for Risk* also provides guidance on risk assessments that enable stakeholders to consider the cost of mitigation and the required resources against the loss exposure [3].

ISACA's 'Securing Mobile Devices' using COBIT 5 for Information Security [3] states that a robust BYOD solution should include security mechanisms with right combination of organizational policies. Understanding the different types of risks existing in BYOD environment will improve the evaluation and assessment of suitable BYOD solution. To support this, the document categorizes risks to include those relating *physical risks, technical risks and organizational risks* [3]. These are subcategorized to include a wide range of risks those related to information transmission on wireless networks, mobility, unencrypted data, authentication, physical risks and ownership issues with device in case of fraud investigation and respective recommendations for addressing these concerns [3]. *Securing Mobile Devices* explains risk scenarios where the system has vulnerabilities that the threat can attack [3]. Explanation of Risk scenarios include classification of smartphone threats in which most of them are applicable for BYOD environment whether it is platform-based or platform-less BYOD solution [3]. With the analysis of risk scenarios presented in this document, an overview of risk involved in implementing BYOD policy at an enterprise is provided. And, the respective risk mitigation techniques recommended in this document explains how to handle the risk appropriately including the requirements and possible breaches that may occur if not handled appropriately. Although this document stays vendor neutral it does not provide a specific approach on selection of platform-based BYOD solution or to develop a platform-less BYOD solution. Also, it does not discuss residual risks that may remain from implementing the recommended risk mitigation techniques.

COBIT 5 for Risk professional guideline categorizes risk events to include those relating to those involved mobile devices. It provides a comprehensive set of generic risk scenarios which includes a wide range of risk scenario categories, components and risk types. This document provides best practices to implement and manage risk mitigation strategies to control risk. This information is useful in effectively managing IT risks in general, but not specific to BYOD environment.

NIST has published a comprehensive guideline for enterprises to manage mobile devices [16]. The guideline highlights high portability of the devices and suggests that enterprises should develop MDM strategies assuming that a device will fall into the hands of malicious users. The document suggests layered security strategy in which initial layer requiring authentication and second layer protecting the information belonging to the enterprise and third layer protecting the most sensitive information. To eliminate related risks, the guideline recommends the strategy to limit access to the devices that are least controlled while most-controlled devices can be given most access such as access to sensitive information. This framework also recommends that the enterprise's BYOD policy should consider risk-based strategy that determines what levels of access to what type of devices should be given. In this guideline, key factors to consider in building BYOD solution such as information sensitivity, level of confidence in security policy compliance, technical limitations for the design phase are explained. Several other factors including methods to dispose decommissioned devices are also reviewed in this NIST guideline [16]. Although best practices and recommendations provide vital information to the enterprises to consider while building the BYOD policy, a guideline to develop a comprehensive platform that can be used to coordinate BYOD related controls is required.

Review of existing industry frameworks suggested that these comprehensive frameworks have intentions to provide generic guidance to all kinds and different sizes of corporations. While these recommendations provide adequate guidance and educate the enterprise GRC function about the best practices to maintain a successful BYOD policy, an approach to evaluate suitable type of BYOD solution is not provided by the frameworks.

Some suggestions have been provided in peer reviewed literature for enterprises on selection of a specific Platform-based BYOD solution [7]. Existing frameworks and peer review documents does not include residual risk and does not identify *control risks* such as risk of incompatibility of different components of the system and risks of reliability of the controls themselves. Although, several frameworks mentioned continuous monitoring and updating will be required to develop a successful solution in the long run, it is not specific about the additional risks that may arise after the risk mitigation techniques have been applied. Therefore, an extended analysis of the risk scenarios, and best practices recommended by the other frameworks and current industry standards is required.

2.2 Review of Strategies Implemented by Existing Platform-Based BYOD Solutions

There are several methodologies that have been developed suggesting different BYOD strategies. Platform-based BYOD solutions are known for providing consistent security policy enforcement throughout the enterprise with alignment of business needs [4]. Such solutions provide BYOD work environment on the devices by restricting the applications that are not developed with a specific framework. This concept is referred as containerization and credibility of an application is determined using black- or white-listing the applications and security mechanisms such as anti-virus tools [4]. Related research on Platform-based BYOD identified the risks involving BYOD program, and researchers have also compared few popular BYOD solutions to help enterprises make a choice depending on the requirements. Also, IT support is a key consideration as it may outweigh what an organization could save on infrastructure, technology and maintenance when contracting with a vendor for Platform-based BYOD solution. These are the factors that appears to be potential benefits in choosing a Platform-based BYOD solution [1]. To understand risk mitigation strategies implemented by Platform-based BYOD solutions, several Platform-based BYOD solutions such as Samsung KNOX and MobileIron have been reviewed to determine how the risk scenarios are handled in them.

Samsung declared KNOX solution as BYOD specific while other vendors have referred their solutions as MDM suites however they claim they serve the purpose of BYOD. It provides security features that enable business and personal content processed on the same device, all in one-touch switching the device from Personal to Work use or vice-versa [2]. KNOX uses National Security Agency's (NSA) patented technologies and hardware-level features to provide improved security for its operating systems and applications [2].

MobileIron is another MDM solution for operating mobile devices in BYOD environment that provides a suggested approach for adopting BYOD policy. To prepare an enterprise for BYOD, MobileIron suggests that the enterprise have to first determine their BYOD risk tolerance by determining device choice, open or restricted organizational policies, and access to applications, maintenance and support. Recommendations say that enterprises has to clearly define BYOD program goals at the initial stage and communication with the end users is important. After identification of their enterprise's IT capabilities and upgrading infrastructure they suggest soft launching the BYOD applications before full phase deployment to fix issues. Regular maintenance and user training is mentioned a key step for successful run of BYOD policy [17].

Platform-based BYOD solutions may be successful in implementation when the requirements of the enterprise suits the vendor-predefined profile. Different types of enterprises have different requirements and different type of functionalities are required to accommodate their requirements. Since the BYOD solution is managed by a vendor, it may not be possible to have enough flexibility to accommodate all kinds of requirements that may arise in future [21]. Application management features may be missing or weak in platform-based BYOD solutions due to limited test environment for company-developed applications [21]. Right mixture of security policies to control devices on the network may not be possible due to lack of flexibility with vendor provided BYOD solution [21]. Addition or removal of certain functionalities based on operations requirement may affect other functions of the BYOD solution to add additional risks.

From the review of existing strategies and existing Platform-based BYOD solutions, it is visible that industrial frameworks and peer reviewed research literature have recommended best practices for adoption of BYOD in generic but not specific to platform-based or platform-less BYOD solutions. Review also revealed that mitigation techniques that have been used by Platform-based BYOD solution providers are quite similar in nature to those strategies recommended in the frameworks with a possibility for some limited customization when required. However, there is no specific set of recommendations that provided enough information to help determine suitable type of BYOD solution i.e. Platform-based or Platform-less BYOD solution.

3 Analysis and Results

To achieve research objectives of comparing platform-based and platform-less BYOD solutions, the study focuses on analyzing the potential ability of the Platform-based and Platform-less BYOD solutions to handle the risks in BYOD environment.

Table 1 shows a sample of analysis of best practices identified from review of frameworks and research literature for each risk area with a possible risk scenarios derived from the framework. Risk mitigation techniques in use by existing platform-based BYOD solutions column explains risk mitigation techniques implemented by current platform-based BYOD solutions for respective risk scenarios as applicable are also outlined in the table. Recommended risk mitigation techniques for BYOD solutions column shows recommendations provided based on the analysis and results of this work, which are not specific to platform-based or platform-less BYOD solutions but suggested strategies that an enterprise can consider to decide between platform-based and platform-less BYOD solutions. Residual risks column presents the results of residual risk analysis performed in this research work, which helps enterprises to foresee potential residual risks or additional risks they will be looking at by adopting platform-based or platform-less BYOD solutions or some combination of them. Full version of Table 1 is available at the following URL: https://sites.google.com/student.concordia.ab.ca/mdm-controls/home.

From Table 1, one can see risk scenarios applicable to BYOD and mitigation strategies suggested by frameworks followed by control strategies knowing to be implemented by existing Platform-based BYOD solution. Recommendations include suggested best practices for the risk scenarios. Based on the analysis, it is clear that platform-based solution operate mainly based on several strategies such as restricting certain applications that are not developed with a specific SDK. For successful run of BYOD policy, an enterprise must take into account not only BYOD risks but also residual risks and control risks related to BYOD as well as some general IT risks such as integration related risks especially for custom made solutions [12].

For example, the fact that BYOD devices are mobile in nature and communicate using wireless networks there is increased amount of risk than that of conventional wired networks. Although, the amount of risk varies from one case to another,

Table 1. Risks and mitigation strategies in MDM environment. Full version is available at the following URL: https://sites.google.com/student.concordia.ab.ca/mdm-controls/home

Risk area	Risk scenario	Best practices recommended by frameworks/peer-reviewed literature	Risk mitigation techniques in use by existing platform-based BYOD solutions	Recommended mitigation techniques for BYOD solutions	Residual risks
Wireless networks [3]	Wireless networks are more vulnerable to attacks thus risk of information interception resulting in security breach is increased. Also, it leads to damage to reputation and failure in adherence to regulations [3]	Defense in depth [17, 18]: Use of multiple security countermeasures in coordination to one another such as hardware restriction mechanisms.	KNOX uses SecureBoot, TrustedBoot, ARM Trustzone – based Integrity Measurement Architecture (TIMA) as a protection [2] Airwatch uses secure content management to protect enterprise data [9]	Use of virtual private networks and deploying containerization engine which is installed atop the host operating system is recommended	Residual risk remains even after data link level encryption and authentication protocols due to wireless nature of communication. Evil twin attack may be launched to steal passwords and sensitive information using a bogus evil twin network which pretends like enterprise's authentic network

recommendations by frameworks and peer-reviewed literature include adoption of defense-in-depth strategy to be considered when designing access control policy [10]. Analysis of existing platform-based BYOD solutions such as Samsung KNOX and Airwatch proved that vendors used SecureBoot, TrustedBoot and TIMA (Techniques of Informatics and Microelectronics for integrated systems Architecture) to protect the device from attacks [9]. With the analysis of best practices and existing BYOD solutions, suggested best practices include use of VPNs and containerization which provides security by isolation. Similarly, key considerations and results of analysis for each scenario as applicable are mentioned in the full version of the table.

Overall, Table 1 can be used to provide recommendations for selecting the type of BYOD solutions along with explaining the potential residual risks that an enterprise can find useful in arriving at a decision. For instance, compatibility issues may not be underestimated or ignored while considering developing a platform-less BYOD solution for an enterprise. Compatibility issues if ignored may lead to potential threat if one of the security controls proves to be incompatible with other components of their BYOD solution. Residual risk analysis for the most common risk scenarios helps enterprises by presenting the significance of the aspect and stresses taking into consideration while developing the BYOD solution.

Table 1 – the main outcome of this research – provides hands-on document for enterprises in general, irrespective of their type. Risk scenarios are developed based on most common issues reported in variety of industries therefore certain adjustments will be required to put suggested controls in place. Analysis on platform-based BYOD policy is limited to two solutions without estimating costs but it gives general understanding of what is available for the enterprises. Although this work does not imply to a specific type of enterprises it has to be said that even with the best BYOD solution in place, organizations that are handling sensitive information may choose to avoid BYOD.

4 Conclusion

In this research risk scenarios and respective risk mitigation techniques previously suggested by both peer reviewed research and globally accepted industry-leading frameworks have been reviewed. Analysis of risk mitigation techniques for the risk scenarios is performed in comparison to the risk mitigation strategies that are being implemented by some BYOD vendors in the market. This comparative analysis along with identification of residual risks serves as the main contribution of this research work. The developed set of recommendations will help enterprises arriving at a decision in identifying a suitable type of BYOD solution for their business needs. Also, residual risks mentioned in Table 1 help enterprises weigh them against the potential reward to determine if it is worth adoption of BYOD policy in the first place. Overall, this work, by providing recommendations based on comparative analysis and residual risk analysis also contributes to the future development of BYOD solutions for enterprise. Recommendations for future work include researching a detailed approach to develop a platform-less BYOD solution. Risk mitigation strategies to control residual risk for platform-based and platform-less BYOD solutions may be developed as a part of future research work.

References

1. Srikant, R., et al.: BYOD in the enterprise - a holistic approach. In: Information Systems Audit and Control Association (ISACA), Available at BYOD-in-the.pdf
2. An Overview of the Samsung KNOX Platform. Samsung Electronics Co. Ltd., Gyeonggi-do, Korea (2013). https://www.samsungknox.com/en/system/files/whitepaper/files/An%20Overview%20of%20the%20Samsung%20KNOX%20Platform_V1.11.pdf
3. Securing Mobile Devices Using COBIT® 5 for Information Security, Information Systems Audit and Control Association (ISACA), Meadows, IL (2012). http://www.isaca.org/Knowledge-Center/Research/Documents/Securing-Mobile-Devices-Using-COBIT-5-for-Information-Security_res_English_1112.pdf
4. Romer, H.: Best practices for BYOD security. Comput. Fraud Secur. **2014**(1), 13–15 (2014)
5. SANS Institute: Securing BYOD with Network Access Control, a Case Study, SANS Institute InfoSec Reading Room (2012). http://www.sans.org/reading-room/whitepapers/access/securing-byod-network-access-control-case-study-34185
6. Gartner: Gartner predicts by 2017, half of employers will require employees to supply their own device for work purposes. Newsroom (2013). Available at d/2466615

7. Odilinye, S., Butakov, S., Kazemeyni, F.: Evaluation criteria for selecting Bring-Your-Own-Device (BYOD) Platform. In: An Enterprise. Decision Sciences Institute. http://www.decisionsciences.org/Portals/16/Proceedings/AM-2014/files/p747282.pdf
8. Arellia: Application control solution datasheet. Application control solution. Whitelisting (2014). http://www.arellia.com/products/application-control-solution/
9. Enabling Bring Your Own Device (BYOD). In: The Enterprise: Leveraging AirWatch to Create a Secure and Convenient BYOD Program, Airwatch LLC (2012). http://www.satisnet.co.uk/pdfs/AirWatch-byod-whitepaper.pdf
10. EY: Bring your own device: Security and risk considerations for your mobile device program. Insights on governance, risk and compliance (2013). http://www.ey.com/Publication/vwLUAssets/EY_-_Bring_your_own_device:_mobile_security_and_risk/$FILE/Bring_your_own_device.pdf
11. Scarfo, A.: New security perspectives around BYOD. In: 2012 Seventh International Conference on Broadband, Wireless Computing, Communication and Applications (BWCCA), pp. 446–451. IEEE, November 2012
12. Gruener: The four BYOD integration challenges. In: BYOD Requirements (2013). http://www.tomsitpro.com/articles/mobility-consumerization-mdm-privacy-it_security,2-506.html
13. Ann: BYOD: (Bring Your Own Device) is your organization ready?. Policy Development (2013). https://www.ipc.on.ca/site_documents/pbd-byod.pdf
14. Samsung. KNOX: KNOX workspace. on-device encryption (2015). https://www.samsungknox.com/en/products/knox-workspace/technical
15. Government Business Council: 3 Mobile security threats a BYOD strategy should prepare for. Attacks to Watch (2013). https://www.f5.com/pdf/analyst-reports/mobility-issue-brief.pdf
16. National Institute of Standards and Technologies (NIST): Guidelines for managing the security of mobile devices in the enterprise. Technologies for Mobile Device Management (2013). http://nvlpubs.nist.gov/nistpubs/SpecialPublications/NIST.SP.800-124r1.pdf
17. MobileIron: The ultimate guide to BYOD. BYOD: driving the mobile enterprise transformation (2015). https://www.mobileiron.com/en/whitepaper/ultimate-guide-byod
18. National Security Agency (NSA): Mobile devices management: a risk discussion for IT decision makers. Threat Model for Enterprise Owned Devices (2012). https://www.nsa.gov/ia/_files/factsheets/mdm_decision_makers.pdf
19. National Institute of Standards and Technologies (NIST): Security for wireless networks and devices (2013). http://www.itl.nist.gov/lab/bulletns/bltnmar03.htm
20. TechTarget: The hidden threat: residual data security risks of PDAs and smartphones (2015). http://searchmHYPERLINK, http://searchmobilecomputing.techtarget.com/feature/The-hidden-threat-Residual-data-security-risks-of-PDAs-and-smartphones
21. Matt, S.: Pros and cons of mobile device management software (2015). http://searchconsumerization.techtarget.com/tip/Pros-and-cons-of-mobile-device-management-software
22. David, N., Esq.: The legal implications of BYOD: preparing personal device use policies. ISSA J., November 2012. http://www.infolawgroup.com/files/2012/12/BYOD_ISSA1112-pdf1.pdf
23. InfoSec Institute: Bringing down security risks with a BYOD encryption policy. BYOD Encryption Challenges (2014). http://resources.infosecinstitute.com/bringing-security-risks-byod-encryption-policy/

An Energy-Efficient Key Agreement Mechanism for Underwater Sensor Networks

Yue Zhao(✉), Bo Tian, Zhouguo Chen, Yiming Liu, and Jianwei Ding

Science and Technology on Communication Security Laboratory, Chengdu 610041, China
yuezhao@foxmail.com, tb_30wish@163.com, czgexcel@163.com, 251508910@qq.com, mathe_007@163.com

Abstract. The key agreement mechanisms are designed according to different communication links of underwater sensor networks. Combined with the underwater nodes deployment and geographic location information, the key agreement mechanism, which using the elliptic curve point multiplication with lower computational overhead, can support the identity authentication and session key agreements between the nodes on the non-bidirectional links, and resist Sybil attacks, replay attacks, spoofed node attacks, node replication attacks, and so on. Extensive simulations demonstrate that the proposed energy-efficient key agreement mechanism has better security and network performance, especially reducing the energy consumption of low-performance sensor nodes. The underwater sensor networks can achieve high network connectivity, meanwhile guaranteeing that the energy consumption of the cluster-head node and sensor node is less than 20 J and 10 J, respectively.

Keywords: Underwater sensor networks · Key agreement · Energy consumption · ID-based · Session key

1 Introduction

It is security as well as supportable network scale that are usually rendered as the design criteria for the traditional key agreement mechanisms. It is hard to maintain underwater nodes, and their communication and computation capacities are also constrained by their own low energy. Therefore, the energy consumption should be taken as an important design criterion of key agreement mechanisms [1, 2]. Considering the characteristics of underwater sensor networks (USNs) and the constrained resources of underwater nodes, the traditional key agreement mechanisms cannot be applied directly to USNs. It is a very challengeable task to build a key agreement mechanism suitable for USNs. Therefore, there is an urgent need for a secure and efficient key agreement mechanism to improve the security of USNs while reducing communication and computational overheads.

Most key agreement mechanisms, in the existing literature, concentrate on wireless sensor networks rather than USNs. An authenticated key agreement mechanism based on self-certified public key is provided in [3], which does not need certificate

K.J. Kim et al. (eds.), *IT Convergence and Security 2017*,
Lecture Notes in Electrical Engineering 450,
DOI 10.1007/978-981-10-6454-8_20

management, thereby suitable for resource-constrained wireless sensor networks. However, it adopts costly Tate pairing operations, so each node has a significant increase in energy consumption and computation time. An identity-based key agreement mechanism is proposed in [4]. Each two nodes establish a session key based on the Diffie-Hellman (DH) key exchange protocol. Although the mechanism has some advantages in energy consumption and computation time of each node, it using the encryption method to identify other nodes will increase the number of interactions and bring higher energy consumption. In [5], a mutual authentication and key exchange mechanism based on a session key is presented to resist replay attack, man-in-the-middle attack and Denial-of-Service (DoS) attack. The mechanism, however, is not suitable for the hierarchical key management mechanism for large and medium-sized wireless sensor networks, since each node needs to preserve the session key with all other nodes. So the mechanism is not suitable for the hierarchical key management architecture for large and medium-sized wireless sensor networks. In summary, the existing key agreement mechanisms are all based on the bilinear mapping or large integer decomposition problems, in which the communication and computation overheads are relatively large. Obviously, the above-mentioned key agreement mechanisms are not applicable to USNs.

This paper proposes an energy-efficient key agreement mechanism for underwater sensor networks. The security analysis and simulation results show that the proposed key agreement mechanism has good security and performance in the computational efficiency and message length of inter-node interactions. The remainder of this paper is organized as follows. Section 2 describes the system model of USNs based on clustering. Section 3 presents the corresponding key agreement mechanisms for the four specific communication links of USNs. Section 4 analyzes the security and efficiency of the key agreement mechanism. Section 5 provides the simulation environment, the numerical results and some discussions. Section 6 concludes this paper.

2 System Model

Figure 1 shows the network model of USNs. In order to reduce the communication overheads among the underwater nodes, USNs are divided into clusters according to some attributes, such as geographical locations [6]. Each cluster has a cluster-head node (H-node), which is responsible for collecting observing data from the sensor nodes, aggregating and sending data to the buoy on water. Each H-node maintains long-term online state, and supports the direct communication with adjacent H-nodes. S-nodes are mainly responsible for observing in local area and sending the observing data to H-nodes. Compared to S-nodes, H-nodes are high-performance nodes with larger storage space as well as stronger computation and communication capacities. Moreover, the number of H-nodes is smaller than that of S-nodes in the observation area.

Each underwater node loads the system parameters, such as identity information, private keys and so on. Each node negotiates with other nodes to obtain the corresponding session keys. According to the identity-based encryption, the private key generator (PKG) first chooses an elliptic curve defined over a finite field, and $\boldsymbol{E}(F_P)$ is a

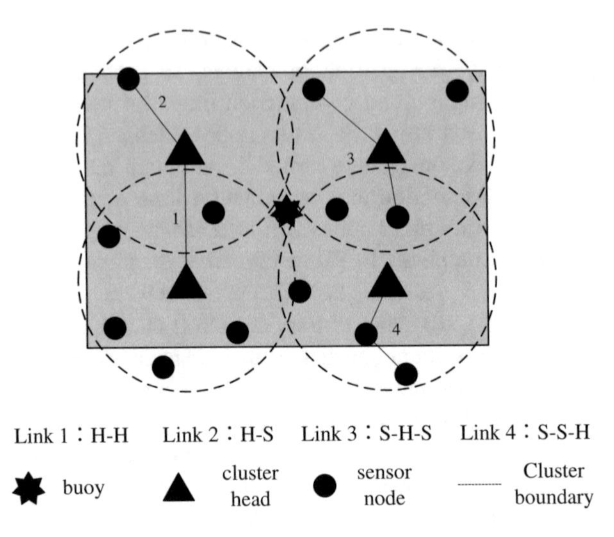

Fig. 1. The network model of USNs

group consisted of points on the elliptic curve and other points at infinity. Given $P \in \boldsymbol{E}(F_P)$, and $\boldsymbol{G}_1$ is a cyclic group generated by P. Next the PKG chooses an integer $s \in \boldsymbol{Z}_q^*$, is a multiplicative group with q non-zero elements in a finite field. Then the PKG calculates its own public key $P_{\text{pub}} = sP$, which is made public to all underwater nodes. Finally, the PKG works with cryptographic hash functions H_1 and H_2, where H_1: $\{0, 1\}^* \times \boldsymbol{G}_1^* \rightarrow \boldsymbol{Z}_q^*$ and H_2: $\{0, 1\}^* \rightarrow \boldsymbol{Z}_q^*$.

Each nodes ID value is unique and different from each other. After being deployed, nodes location remains the same in general. When performing the localization algorithm in the clustering process, each node generates its own public key and private key with identity information and geography information. Take a H-node H_A for example, the public key based on identity information ID_A and geography information l_A is calculated by $c_A = H_1(ID_A \| l_A \| R_A)$, and the private key expressed as (R_A, d_A), where $R_A = r_A P$, $r_A \in \boldsymbol{Z}_q^*$, and $d_A = r_A + c_A s \bmod q$.

It is worth noting that the difference in communication distance and the energy consumption between H-nodes and S-nodes leads to the existence of the non-bidirectional links. It means two underwater nodes cannot communicate with each other directly, even though they are in the same cluster geographically. Therefore, the key agreement mechanism needs to take the existence of non-bidirectional links into consideration. It is critical not only to ensure non-adjacent S-nodes can complete key agreement through H-node, but also the remote S-node can establish the session key with the H-node to realize the end-to-end secure communication as well.

Therefore, the key agreement mechanisms for USNs include four typical cases as shown in Fig. 1. First, the key agreement on bidirectional link between an H-node and another adjacent H-node (H-H); second, on bidirectional link between an H-node and an adjacent S-node (H-S); third, on non-bidirectional link between two non-adjacent S-nodes (S-H-S); fourth, on non-bidirectional link between an H-node and a remote S-node in the same cluster (S-S-H).

3 Authentication and Key Agreement

3.1 H-H Mechanism

If two H-nodes H_A and H_B are within the each other's communication range, there is a bi-directional communication link between H_A and H_B. With high performance, H-nodes can use different session keys in each session or periodically negotiate different session keys. The authentication and key agreement mechanism between H_A and H_B are presented in Fig. 2.

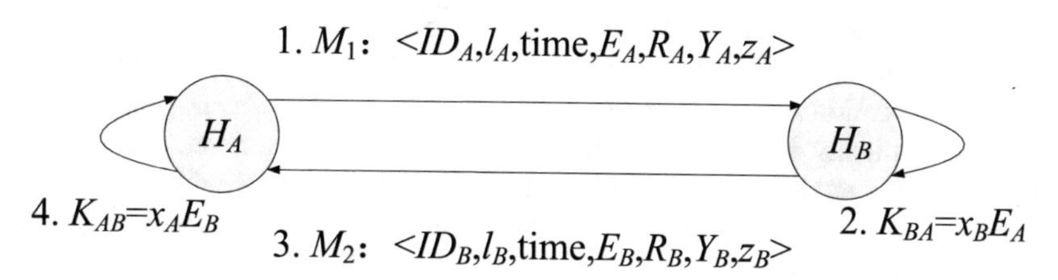

Fig. 2. The key agreement mechanism between two H-nodes

1. Given x_A, $y_A \in Z_q^*$, H_A computes $Y_A = y_AP$, $E_A = x_AP$, and authentication information $h_A = H_2(ID_A, l_A, time, E_A, R_A, Y_A)$, where *time* is represented as a timestamp, which is used to determine the time of sending or receiving messages under the premise of keeping time synchronization among underwater nodes. Then H_A calculates $z_A = x_A + h_Ad_A\text{mod}q$, and sends M_1:<ID_A, l_A, *time*, E_A, R_A, Y_A, z_A> to H_B.
2. When message M_1 is received, H_B first determines whether time is valid, obtains identity and geographic information, ID_A and l_A, and then verifies the validity of H_A identity through the public key c_A and the authentication information h_A, i.e.

$$z_AP = (y_A + h_Ad_A)P = Y_A + h_A(r_A + c_As)P = Y_A + h_A\left(R + c_AP_{pub}\right) \tag{1}$$

If the above information is verified, the H_A identity is legal. Otherwise, the message M_1 is rejected. Given x_B, $y_B \in Z_q^*$, H_B computes Y_B, E_B, h_B, z_B, and K_{BA} as

$$\begin{cases} Y_B = y_BP, \\ E_B = x_BP, \\ h_B = H_2(ID_B, l_B, time, E_B, R_B, Y_B), \\ z_B = x_B + h_Bd_B \bmod q, \\ K_{BA} = x_BE_A. \end{cases} \tag{2}$$

3. H_B sends M_2:<ID_B, l_B, *time*, E_B, R_B, Y_B, z_B> to H_A.
4. When receiving the message M_2, H_A first determines whether M_2 is an expired message, and then verifies the H_B identity through the public key c_B and the authentication information h_B. If $z_BP = Y_B + h_B(R_B + c_BP_{\text{pub}})$ is workable, H_A continue to calculate $K_{AB} = x_AE_B$.

So far H_A and H_B get the session key as

$$SK = H_2(ID_A, ID_B, l_A, l_B, K_{AB}) = H_2(ID_A, ID_B, l_A, l_B, K_{BA}). \tag{3}$$

3.2 H-S Mechanism

In USNs, S-nodes need to continually send underwater observation information to H-nodes. These interactions cause frequent communication and a vast amount of data. Because the performance of the S-node is low, it is uneasy to adopt the security communication mode of one-time pad for the interaction between the S-node and H-node. The S-node and H-node can establish the session key when implementing mutual authentication and session key agreement in the deployment phase of USNs, and then periodically implement re-authentication and key agreement to update the session key. The session key agreement of H-S mechanism is similar to that of H-H mechanism. H_A and S_B can obtain the session key $SK = H_2(ID_A, ID_B, l_A, l_B, K_{AB}) = H_2(ID_A, ID_B, l_A, l_B, K_{BA})$.

3.3 S-H-S Mechanism

In the deployment phase of USNs, each S-node has established a session key with the H-node in the same cluster. Thereafter, non-adjacent S-nodes can establish and periodically update session keys with the assistance of the H-node. For example, two S-nodes, S_A and S_B, establish the session keys SK_{AC} and SK_{BC} respectively with the H-node H_C according to the H-S mechanism. The key agreement mechanism in the participation of H_C between S_A and S_B is shown in Fig. 3.

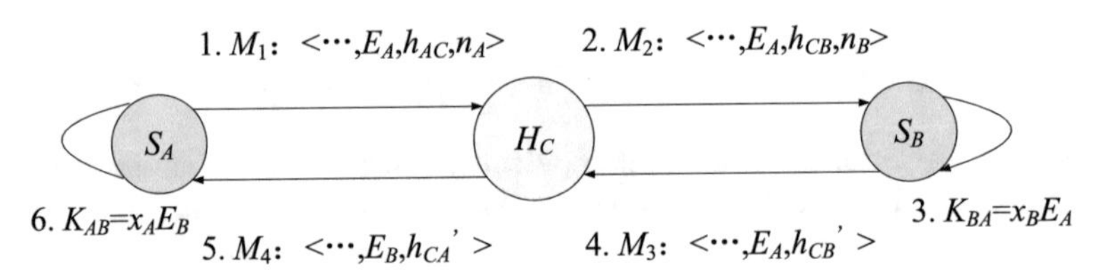

Fig. 3. The key agreement mechanism between two non-adjacent S-nodes

1. Choosing $x_A, n_A \in \mathbf{Z}_q^*$, S_A computes $E_A = x_A P$, $h_{AC} = H_{SK_{AC}}(ID_A, ID_B, l_A, time, E_A, n_A)$, and sends M_1:$<ID_A, ID_B, l_A, time, E_A, h_{AC}, n_A>$ to H_C.
2. When receiving the message M_2, H_C determines whether *time* is valid, and computes $h_{CA} = H_{SK_{CA}}(ID_A, ID_B, l_A, time, E_A, n_A)$.

If h_{AC} and h_{CA} are equal, S_A will be authenticated by H_C. Given $n_B \in \mathbf{Z}_q^*$, H_B computes $h_{CB} = H_{SK_{BC}}(ID_C, ID_A, l_C, l_A, time, E_A, n_B)$, and sends M_2:$<ID_C, ID_A, l_C, l_A, time, E_A, h_{CB}, n_B>$ to S_B.

3. S_B verifies the timeliness of M_2, calculates $h_{BC} = H_{SK_{BC}}$ (ID_C, ID_A, l_C, l_A, $time$, E_A, n_B), and determines whether h_{BC} and h_{CB} are equal. If the above conditions are true, S_A and H_C are authenticated by S_B. Given $x_B \in \boldsymbol{Z}_q^*$, S_B computes E_A, K_{BA}, and h'_{BC} as

$$\begin{cases} E_A = x_A P, \\ K_{BA} = x_B E_A, \\ h'_{BC} = H_{SK_{BC}}(ID_B, ID_A, l_B, time, E_B, n_B). \end{cases} \quad (4)$$

4. S_B sends M_3:<ID_B, ID_A, l_B, $time$, E_B, h'_{BC}> to H_C.
5. H_C determines whether M_3 is valid, and computes $h'_{CB} = H_{SK_{CB}}$(ID_B, ID_A, l_B, $time$, E_B, n_B).

If h'_{BC} and h'_{CB} are equal, S_B will be authenticated by H_C. H_C then computes $h'_{CA} = H_{SK_{CA}}$(ID_C, ID_B, l_C, l_B, $time$, E_B, n_A), and sends M_4:<ID_C, ID_B, l_C, l_B, $time$, E_B, h'_{CA}> to S_A.

6. S_A determines the timeliness of M_4, and computes $h'_{AC} = H_{SK_{AC}}$ (ID_C, ID_B, l_C, l_B, $time$, E_B, n_A). If $h'_{CA}$and h'_{CA} are equal, H_C and S_B will be authenticated successfully by S_A. Finally, S_A computes $K_{AB} = x_A E_B$.

Two non-adjacent S-nodes S_A and S_B obtain the session key SK as

$$SK = H_2(ID_A, ID_B, l_A, l_B, K_{AB}) = H_2(ID_A, ID_B, l_A, l_B, K_{BA}).$$

3.4 S-S-H Mechanism

In USNs, most of S-nodes have bi-directional communication links with H-nodes in the same cluster, while few non-bidirectional links exit between S-nodes and H-nodes. In order to improve the network connectivity, it is necessary to implement authentication and key agreement for the non-bidirectional links. Assuming that there is a non-bidirectional link between the S-node S_A and the h-node H_B. The S-node S_C is adjacent to both S_A and H_B. H_B establishes the session key SK_{BC} with the S_C according to the H-S mechanism.

In Fig. 4, S_A and H_B on a non-bidirectional link implement authentication and key agreement with the participation of S_C. The process of establishing a session key between S_A and S_C is similar to that of the H-H mechanism. S_C, as a relay node, assists S_A and H_B to complete the authentication and key agreement.

1. Given x_A, $n_A \in \boldsymbol{Z}_q^*$, S_A computes $E_A = x_A P$, $h_{AC} = H_{SK_{AC}}$(ID_A, ID_B, l_A, $time$, E_A, n_A), and sends M_1:<ID_A, ID_B, l_A, $time$, E_A, h_{AC}, n_A> to S_C.
2. S_C determines the timeliness of M_1, and computes $h_{CA} = H_{SK_{CA}}$(ID_A, ID_B, l_C, l_B, $time$, E_A, n_A). If h_{AC} and h_{CA} are equal, S_A is authenticated by H_C. Given $n_B \in \boldsymbol{Z}_q^*$, S_C computes $h_{CB} = H_{SK_{CB}}$(ID_C, ID_A, l_C, l_A, $time$, E_A, n_B), and sends M_2:<ID_C, ID_A, l_C, l_A, $time$, E_A, h_{CB}, n_B> to H_B.

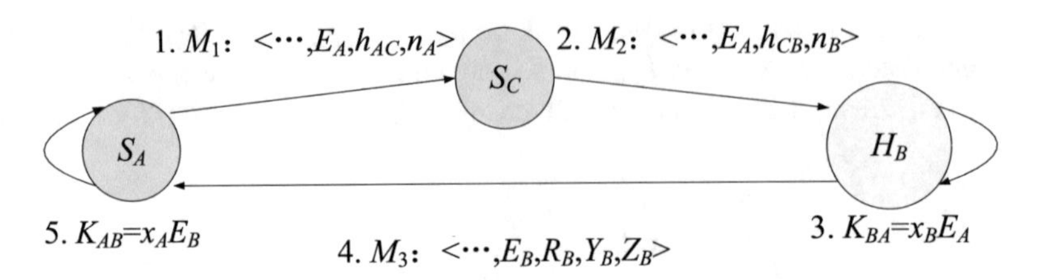

Fig. 4. The key agreement mechanism between the H-node and the S-node on a non-bidirectional link

3. H_B verifies the timeliness of M_2, calculates $h_{BC} = H_{SK_{BC}}(ID_C, ID_A, l_C, l_B, time, E_A, n_B)$, and determines whether h_{BC} and h_{CB} are equal. If the above conditions are true, S_A and S_C are authenticated by H_B. Given $x_B, y_B \in \mathbf{Z}_q^*$, H_B computes Y_B, E_B, h_B, z_B and K_{BA} as

$$\begin{cases} Y_B = y_BP, \\ E_B = x_BP, \\ h_B = H_2(ID_B, l_B, time, E_B, R_B, Y_B), \\ z_B = y_B + h_Bd_B \bmod q, \\ K_{BA} = x_BE_A. \end{cases} \tag{5}$$

4. H_B sends M_3:<ID_A, ID_B, l_B, $time$, E_B, R_B, Y_B, z_B> to S_A.
5. S_A determines whether M_3 is valid, computes $c_B = H_1(ID_B||l_B||R_B)$, $h_B = H_2(ID_B, l_B, time, E_B, R_B, Y_B)$, verifies the H_B identity, i.e.

$$z_BP = (y_B + h_Bd_B)P = Y_B + h_B(r_B + c_Bs)P = Y_B + h_B(R_B + c_BP_{pub}), \tag{6}$$

and then continue to calculate $K_{AB} = x_AE_B$.

As yet, the H-node and S-node on the non-bidirectional link get the session key $SK_{AB} = H_2(ID_A, ID_B, l_A, l_B, K_{AB}) = H_2(ID_A, ID_B, l_A, l_B, K_{BA})$.

4 Security and Efficiency Analysis

4.1 Security Analysis

1. The proposed key agreement mechanisms realize the forward security [7]. Even if the attacker obtains the master key of the PKG or the private keys of underwater nodes, the session key cannot be obtained by mathematical derivation. The above four mechanisms adopt the elliptic curve digital signature algorithm to achieve identity authentication and obtain session keys. As long as the elliptic curve discrete logarithmic difficult problem is not cracked, the attacker cannot capture x_A and x_B from the message $E_A = x_AP$ and $E_B = x_BP$ exchanged in the key agreement process, nor can x_Ax_BP be calculated. Since the session key is determined only by the

random number $x_A x_B$, the security of the previously established session key is not affected even if the private keys of user A and B, as well as the system master key is leaking. Therefore, this key agreement mechanism has perfect forward security.

2. The proposed key agreement mechanism can resist replay attacks and spoofed node attacks. Timestamps are added to the authentication information to prevent replay attacks. The HH and H-S mechanisms adopt the digital signature algorithm to achieve a two-way identity authentication. The two sides of communication use their own private key to sign the key exchange information in order to facilitate the identification of key exchange information. The attacker cannot know the private keys of participants from both sides, and is unable to forge signatures to pass identity authentication.
   The S-S-H mechanism is based on the H-H and H-S mechanism. The two S-nodes use their session keys that established with the H-node and authentication information to achieve identity authentication. Since the attacker does not establish session keys with the H-node, he cannot complete the authentication and session key agreement process with S-nodes through the H-node. Similarly, the S-H-S mechanism prevents unauthorized users from passing authentication and establishing session keys. Therefore, the four mechanisms can effectively prevent forged attacks.
3. The proposed key agreement mechanism can resist Sybil attacks [8]. The public keys, private keys and authentication messages of all underwater nodes contain identity and geographic information. Since the attacker does not acquaint the private key of each node, he cannot be disguised as a number of nodes with different identities or geographical locations.
4. The proposed key agreement mechanism can resist node replication attacks [9]. The attacker captures an underwater node and places its copies in some geographical locations. Since the four mechanisms use the adjacent node authentication algorithm, every legitimate node outside the surrounding area of the captured nodes will refuse to receive the information from the replicated node. The replicated node cannot authenticate with other nodes. The attacker is not able to carry out attacks by copying information from legitimate nodes.

4.2 Efficiency Analysis

Table 1 shows the computational and communication overhead of the H-H, H-S, S-H-S, and S-S-H mechanism respectively. The computational overhead is the amount of computation in the key agreement process of each node. The communication overhead is the length of messages sent and received by each node in a key agreement process. In Table 1, S denotes an elliptic curve point multiplication operation, M a modular multiplication operation, and H a one-way Hash operation. Assuming that the energy consumed in performing an elliptic curve point multiplication operation is about 0.752 mJ. The energy consumed in transmitting or receiving a byte is about 0.002 mJ. The length of message authentication code C, including E, R, Y, z, n, h, etc., is 20 Byte. The length of *time*, *ID*, and l is 2 Byte, 2 Byte, and 4 Byte, respectively [10].

As can be seen from Table 1, the computational and communication overhead of the H-node is much larger than that of the S-node. The energy consumption of the

Table 1. The computational and communication overhead of four key agreement mechanisms

Key agreement	Mechanism	Computational overhead	Communication overhead
H-H	H-node1	$4S + 2M + 3H$	$4\|C\|+2\|time\|+\|ID\|+\|l\|$
	H-node2	$4S + 2M + 3H$	$4\|C\|+2\|time\|+\|ID\|+\|l\|$
H-S	H-node	$4S + 2M + 3H$	$4\|C\|+2\|time\|+\|ID\|+\|l\|$
	S-node	$4S + 2M + 3H$	$4\|C\|+2\|time\|+\|ID\|+\|l\|$
S-H-S	S-node1	$2S + 2H$	$3\|C\|+\|time\|+2\|ID\|+\|l\|$
	H-node	$4H$	$5\|C\|+2\|time\|+4\|ID\|+4\|l\|$
	S-node2	$2S + 2H$	$2\|C\|+\|time\|+2\|ID\|+\|l\|$
S-S-H	S-node1	$3S + 2H$	$3\|C\|+\|time\|+2\|ID\|+\|l\|$
	S-node2	$2H$	$3\|C\|+2\|time\|+2\|ID\|+2\|l\|$
	H-node	$3S + 2H$	$4\|C\|+\|time\|+2\|ID\|+\|l\|$

H-node is about 12.146 mJ according to the S-H-S mechanism, which is the largest energy consumption in four mechanisms. This level of energy consumption is acceptable for high performance nodes.

5 Performance Evaluation

The system-level simulation parameters are shown in Table 2. The underwater nodes are randomly deployed in the scene of $200 \times 200 \times 100\ m^3$. There is a bidirectional link between each two H-nodes. The experimental results do not consider boundary effects among clusters and the subsidence of underwater nodes.

Table 2. System-level simulation parameters

	Parameter	Value
H-node	Number of nodes	4
	Transmission distance	100 m
	Underwater transmission rate	19.2 kbit/s
	Working voltage	3 V
	Working current	66 mA
	Duration of Tate pairing operation [11]	0.06 s
	Duration of elliptic curve point multiplication operation	0.012 s
S-node	Number of nodes	160
	Transmission distance	30 m
	Underwater transmission rate	4.8 kbit/s
	Working voltage	0.95 V
	Current (working state/receiving state/transmission state) [12]	8/10/27 mA
	Duration of Tate pairing operation	2.66 s
	Duration of elliptic curve point multiplication operation	0.81 s

Figure 5 shows the comparisons of the S-node energy consumption in S-H-S mechanism and other existing key agreement mechanisms. In the identity-based key agreement mechanism proposed in [4], the S-node directly uses the Tate pairing algorithm to authenticate and establish session keys with the H-node. The mechanism brings the most energy consumption in the three mechanisms illustrated in Fig. 5. In particular, when the number of nodes adjacent to the S-node reaches 40, the energy consumption of the S-node exceeds 30 J. In the temporal-credential-based key agreement mechanism proposed in [5] and the S-H-S mechanism, the energy consumption of the S-node is less than 10 J. The S-H-S mechanism is a key agreement mechanism based on public key cryptography. S-node can periodically update session keys with adjacent nodes. Therefore, the S-H-S mechanism has higher security than the temporal-credential-based key agreement mechanism, and the energy consumption of the S-node does not increase significantly as the number of adjacent nodes increases. As can be seen from Fig. 5, the energy consumption of the S-node performing key agreement is generally lower than 5 J.

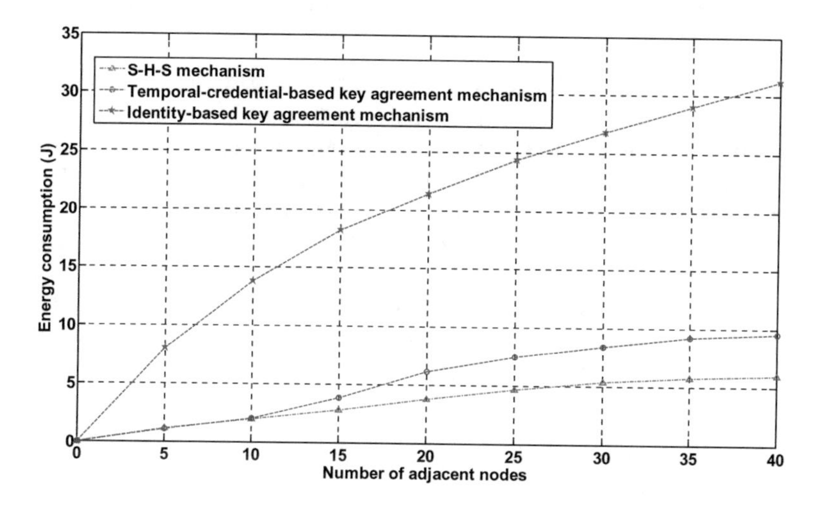

Fig. 5. Energy consumption of the S-node among different mechanisms

Figure 6 presents the energy consumption of the H-node in the S-H-S mechanism. It can be seen that the number of communication links within the cluster will affect the energy consumption of the H-node. Assuming that the number of S-nodes adjacent to the H-node is x, and there are k communication links between S-nodes. The minimum number of communication links between S-nodes is calculated as $k_{min} = \lceil x/2 \rceil$, and the maximum number of communication links is calculated as $k_{max} = x(x-1)/2$. As shown in Fig. 6, when $k = k_{max}$ and $x = 30$, the energy consumption of the H-node is about 20 J, which is acceptable for high-performance H-nodes. In practical applications, the number of communication links between S-nodes is less than k_{max} to achieve a high network connectivity rate, so it can also support the H-node to access a larger number of S-nodes. For the two common scenarios, $k = 1/4k_{max}$ and $k = 1/2k_{max}$, the

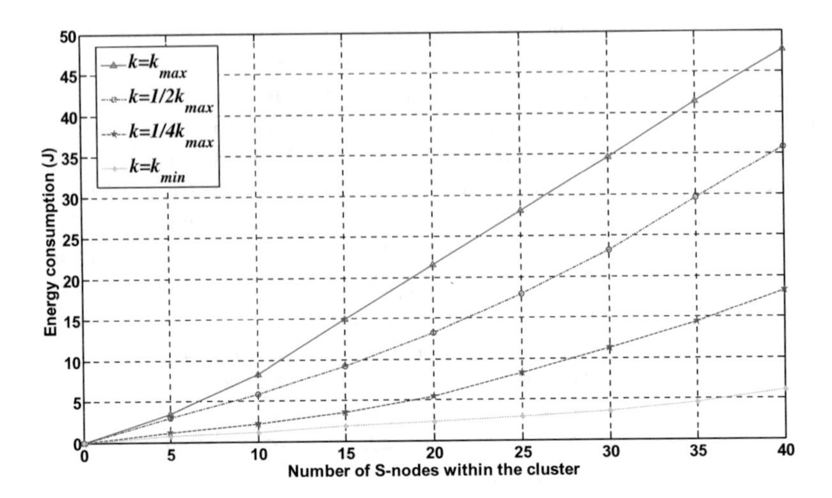

Fig. 6. Energy consumption of the H-node with different communication links

energy consumption of the H-node is generally less than 20 J when $x \leq 27$. Analysis and simulation results show that the completion of S-node authentication and key agreement with the help of the H-node can significantly reduce the energy consumption without causing a large load to the H-node. The session keys between S-nodes can be periodically updated, thereby enhancing the security of USNs.

Figure 7 shows the energy consumption of the nodes according to the S-S-H mechanism. S-node 1 performs key agreements with the H-node on the non-bidirectional link, requiring S-node 2 assistance in the middle. If an S-node helps too many nodes on non-bidirectional links, it will consume too much of its own energy. Therefore, the S-nodes with higher residual energy are often selected as helpers for key

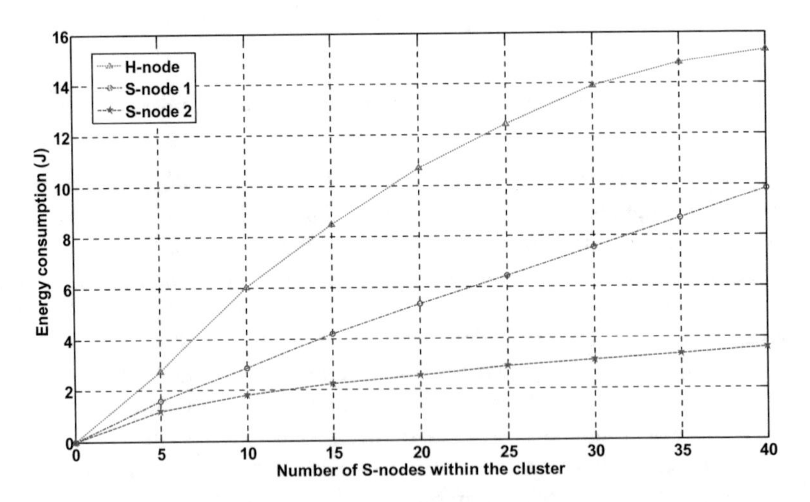

Fig. 7. Energy consumption of the nodes in the S-S-H mechanism

agreements. As shown in Fig. 7, the energy consumption of the H-node is higher than that of S-node 1 and S-node 2, while S-node 2 has the lowest energy consumption. When the number of adjacent nodes is 40, the highest energy consumption of S-node 2 is about 3.6 J. The low performance S-nodes are fully capable of collaborating for key agreements. In addition, because the number of nodes on non-bidirectional links in USNs is small, the S-S-H mechanism will not cause significant energy consumption to S-nodes.

6 Conclusion

In this paper, an energy-efficient key agreement mechanism for USNs is designed according to the resource-constrained characteristics of underwater nodes. In the novel key agreement mechanism, the ability to resist attacks is enhanced by adding identity and geographic information to the public key and private key of underwater nodes. The mechanism does not adopt the complex Tate pairing or large integer decomposition problems. Thus it has a lower computational overhead. The high-performance nodes assist low-performance nodes by means of identity authentication and session key agreement operations, and undertake more auxiliary communication and computing tasks. It effectively reduces the energy consumption of low-performance nodes. A different session key can be negotiated before each communication between cluster head nodes. The session key of the low-performance S-node can be updated periodically to improve the security and robustness of USNs. The analysis and simulation results show that the novel mechanism has better security and network performance, and can be applied to USNs better, in which reduces the communication and computational overhead of underwater nodes. It also provides available references to building security protection mechanism for USNs.

References

1. Cong, Y., Yang, G., et al.: Security in underwater sensor network. In: Proceedings of the International Conference on Communications and Mobile Computing, pp. 162–168, April 2010
2. Li, H., He, Y., et al.: Security and privacy in localization for underwater sensor networks. IEEE Commun. Mag. **53**(11), 56–62 (2015)
3. Ren, Y., Wang, J., et al.: Key agreement protocol for wireless sensor networks using self-certified public key system. J. Comput. Res. Dev. **49**(2), 304–311 (2012)
4. Ghoreishi, S., Razak, S., et al.: New secure identity-based and certificateless authenticated Key Agreement protocols without pairings. In: Proceedings of the International Symposium on Biometrics and Security Technologies, pp. 188–192, August 2014
5. Xue, K., Ma, C., et al.: A temporal-credential-based mutual authentication and key agreement scheme for wireless sensor networks. J. Netw. Comput. Appl. **36**(1), 316–323 (2013)
6. Han, G., Jiang, J., et al.: A survey on mobile anchor node assisted localization in wireless sensor networks. IEEE Commun. Surv. Tutor. **18**(3), 2220–2243 (2016)

7. Wei, Z., Yang, G., et al.: Security of underwater sensor networks. Chin. J. Comput. **35**(8), 1594–1606 (2012)
8. Moradi, S., Alavi, M.: A distributed method based on mobile agent to detect Sybil attacks in wireless sensor networks. In: Proceedings of the International Conference on Information and Knowledge Technology, pp. 276–280, September 2016
9. Dimitriou, T., Alrashed, E., et al.: Imposter detection for replication attacks in mobile sensor networks. Comput. Netw. **108**(24), 210–222 (2016)
10. Xu, J., Song, Y., et al.: Underwater wireless transmission of high-speed QAM-OFDM signals using a compact red-light laser. Opt. Express **24**(8), 8097–8109 (2016)
11. Ramesh, C., Rao, K., et al.: Identity-based cryptosystem based on tate pairing. Glob. J. Comput. Sci. Technol. **16**(5), 93–105 (2016)
12. Pham, C.: Communication performances of IEEE 802.15.4 wireless sensor motes for data-intensive applications: a comparison of WaspMote, Arduino MEGA, TelosB, MicaZ and iMote2 for image surveillance. J. Netw. Comput. Appl. **46**(8), 48–59 (2014)

Simulation Study of Single Quantum Channel BB84 Quantum Key Distribution

Oi-Mean Foong(✉), Tang Jung Low, and Kah Wing Hong

Computer and Information Sciences Department,
Universiti Teknologi PETRONAS, Bandar Seri Iskandar,
31750 Tronoh, Malaysia
{foongoimean, lowtanjung}@utp.edu.my, h.
kahwing@outlook.com

Abstract. With the increasing information being shared online, the vast potential for cybercrime is a serious issue for individuals and businesses. Quantum key distribution (QKD) provides a way for distribution of secure key between two communicating parties. However, the current Quantum Key Distribution method, BB84 protocol, is prone to several weaknesses. These are Photon-Number-Splitting (PNS) attack, high Quantum Bit Error Rate (QBER), and low raw key efficiency. Thus, the objectives of this paper are to investigate the impacts of BB84 protocol towards QBER and raw key efficiencies in single quantum channel. Experiments were set up using a QKD simulator that was developed in Java NetBeans. The simulation study has reaffirmed the results of QBER and raw key efficiencies for the single quantum channel BB84 protocol.

Keywords: Quantum Key Distribution · BB84 · Cryptosystem · QBER · Raw key efficiencies

1 Introduction

Physicists and computer scientists are working together in pursuing the construction and realization of quantum computer, which would exploit and harness the quirks of quantum mechanics to perform certain computations that are much more efficient and scalable fault-tolerant quantum computers [1, 2].

From years of research, quantum computers are found to be theoretically more powerful than conventional computers for breaking the cryptographic codes that are currently used for many aspects particularly monetary transactions and maintaining information confidentiality on the World Wide Web [3]. The security of the current encryption method which is the public key ciphers are based on mathematical calculations that are simple to compute but require an infeasible amount of processing power and time to invert [4]. The current encryption methods are thus facing multiple challenges from threats due to weak random number generators, advances to computational power, new attack strategies and the emergence of quantum computers.

Bennett and Brassard [5] devised the first Quantum Key Distribution protocol in 1984, known as the BB84 protocol. BB84 protocol utilizes the principle of uncertainty formulated by Stephen Wiesner in 1969 [6] and No-Cloning Theorem. However, even though BB84 protocol provides theoretically unconditional security, it is prone to

K.J. Kim et al. (eds.), *IT Convergence and Security 2017*,
Lecture Notes in Electrical Engineering 450,
DOI 10.1007/978-981-10-6454-8_21

several weaknesses. These are Photon Number Splitting (PNS) attacks [7], high quantum bit error rate (QBER) and low raw key efficiency [8, 9].

The objective of this work is therefore to improve the average quantum bit error rate and raw key efficiency. By minimizing the average QBER in the process of Quantum Key Distribution, an eavesdropper will have a minimal amount of information and knowledge about the secret key, effectively ensuring QBER to be below the threshold governed by the two communicating parties (sender and receiver). By improving raw key efficiency in the process of Quantum Key Distribution, more bits generated by the sender will be used as the final secret key between the two communicating parties. Section 2 analyzes the literature review, Sect. 3 introduces the single quantum channel model, Sect. 4 discusses the simulation results based on BB84 protocol, and lastly, Sect. 5 concludes this paper.

2 Literature Review

This section discusses the fundamental theories of Quantum Mechanics, the Quantum Bit Error Rate and Raw Key Efficiency that ensure the confidentiality of the transmitted key during secret key transmission.

2.1 Fundamental Theories of Quantum Mechanics

Quantum Key Distribution exploits the strange rules of quantum mechanics to achieve unconditional security for two communicating parties. These counter-intuitive behaviors and properties of quantum mechanics have direct consequences and pose a challenge in the field of quantum cryptography.

In quantum computing, a quantum bit or qubit is the basic unit of quantum information. It is the quantum analogue to the classical bits in conventional computers. Unlike classical bits which can only represent 1 or 0, quantum mechanics allows the qubit to be in a superposition of both states, thus representing both 1 and 0 at the same time [10]. As a result, the qubit can be represented as a linear combination of $|1\rangle$ and $|0\rangle$.

$$|\varphi\rangle = \alpha|1\rangle + \beta|0\rangle \tag{1}$$

where α and β are probability amplitudes and can be complex numbers. When the qubit is measured in the standard basis, the probability of outcome $|1\rangle$ is $|\beta|^2$ and the probability of outcome $|0\rangle$ is $|\alpha|^2$ where

$$|\alpha|^2 + |\beta|^2 = 1 \tag{2}$$

In the application of a Quantum Key Distribution protocol, a photon, which is light energy, may be used to represent a qubit for information encoding [11]. Apart from that, quantum states present in a quantum system are very frail and unintended measurement as well as exposure to the external environment will destroy its superposition state. This makes qubit in the quantum system difficult to be controlled and manipulate. This quantum behavior is described as quantum decoherence and it provides an advantage in detecting eavesdropper in an event of key sharing in QKD. Therefore, qubit used in key distribution must be well secluded from the environment to ensure the quantum states are well protected [12].

According to Oraevsky [13], when a quantum computer contains more than one qubit, measurement of states independently is generally not possible as the qubits can be linked together in such a way that their states become interdependent. This is known as quantum entanglement and qubits may remain correlated to each other regardless of their distances apart. Quantum entanglement can be utilized in QKD where one photon from each pair will be distributed between the sender and receiver to produce a tentative secret key.

First discovered by Wooters and Zurek [14], the Quantum No-Cloning Theorem effectively ensures that a qubit of unknown state cannot be replicated, duplicated or cloned. It is a protection mechanism in quantum theory and is one of the essential concepts used in quantum cryptography, specifically in QKD. Copies of quantum states are therefore cannot be obtained and an eavesdropper cannot replicate quantum information.

The QKD relies on the Heisenberg Uncertainty Principle articulated by Werner Heisenberg [15], a German physicist who states that two properties of an object cannot be measured at the same time, even in theory. For instance, if a velocity of an atom is to be measured, the position of the atom cannot be accurately determined. Thus, in QKD application, the polarization state of a photon cannot be measured simultaneously by multiple polarizers without randomizing either of the measurements, resulting in inaccurate results [16].

2.2 Quantum Bit Error Rate (QBER)

The Quantum Bit Error Rate (QBER) is the measurement of the probability or percentage of error across the quantum channel during key distribution [10]. It is one of the key quantities in QKD to evaluate the quality of light transmission in QKD system. The quality of signal transmission can be affected by numerous factors including the presence of an eavesdropper, the protocol used, disturbance and noise due to the imperfection of components and transmission impairments to the qubits. In simpler terms, QBER is defined as the ratio of error rate to the key rate and contains information in the presence of eavesdropper and how much this information was compromised. The QBER can be calculated as:

$$QBER = \frac{E}{B_{max}} \quad (3)$$

where E is the length of erroneous bits and B_{max} is the length of the raw key.

In the ideal quantum channel, QBER is equal to zero and any interference by an eavesdropper will result in an increase in QBER. However, in the practical world, noises and disturbance due to external sources have to be taken into account. Hence, a QBER threshold has to be set to determine the level of privacy for the transmitted key.

2.3 Raw Key Efficiencies

Many experimental and even commercial QKD systems based on the BB84 protocol have been proposed, which mainly used photons to represent qubits. An inherent weakness of such systems is that the photons can be easily absorbed by the channel thus only a small percentage of these photons reach the receiver, the majority will be lost in the channel [9]. The qubits received are therefore precious and should not be wasted. In the BB84 scheme, both the sender and receiver may use different basis with 50% probability in which case at least half of the received qubits will be discarded, reducing the efficiency of the system significantly. The raw key efficiency can be calculated using Eq. 4.

$$F = \frac{B_{max}}{R} \tag{4}$$

where $B_{\max}$ is the length of raw key and R is the length of the original random key.

2.4 Recent Work

In a 2015, Li et. al. proved the importance of randomness in QKD and that the security of the final secret key will be compromised if some random input bits are known or controlled by an eavesdropper during the process of key transmission. The security of the BB84 protocol was analyzed against the strong randomness attack where some of the random input bits were completely controlled by an eavesdropper. In weak randomness attack analysis, the random input bits were partially controlled by an eavesdropper. Mogos [18] implemented QKD protocol which utilized a four-state system with twelve orthogonal states. In the simulation, the average Quantum Ququarts Error Rate (QQqER) of the four-state system is found to be 68.34% with the absence of an attacker, and 88.94% with the presence of an attacker. Similarly, Senekane et al. [19] demonstrated an optical implementation of three-non-orthogonal states and a total of six states for a QKD protocol where the proposed protocol achieved a higher security margin, heightened eavesdropper detections and an improved symmetry. In 2016, Mafu [20] investigated the security proof for QKD protocol which utilizes the quantum entanglement. A simple security proof was proposed where an eavesdropper can only guess the output state with a probability that will ensure that the eavesdropper may not obtain more than half of the classical Shannon information regardless of the state transmitted by the sender.

3 Single Quantum Channel Model

This section discusses the existing single quantum channel key distribution model, known as the BB84 protocol.

3.1 BB84 Protocol Model

In BB84 protocol, both sender and receiver, conventionally named Alice and Bob, will each possess a dedicated QKD device with a single quantum channel capable of transmitting polarized photon and a classical channel for bases comparison and transmission of encrypted text.

Subsequently, an additional quantum channel will be proposed in next phase of this research work and integrated into the single Quantum Key Distribution Model as a new method to improve QBER and raw key efficiencies as shown in Fig. 1.

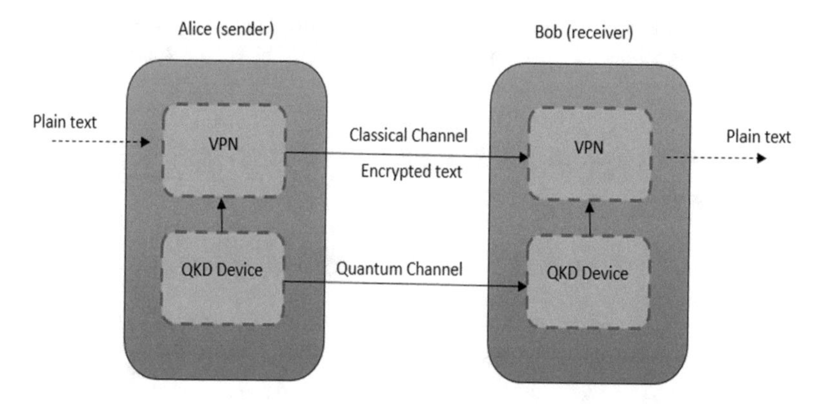

Fig. 1. Single-QKD model and the proposed quantum channel

3.2 BB84 Algorithm

The BB84 protocol includes two bases i.e. rectilinear base and diagonal base while the polarization state of a photon is used to represent each bit value. The 90° and 135° polarization states are used to represent the binary 1 while 0° and 45° polarization states are used to represent the binary 0. The essential secret key sharing phases of the BB84 protocol pseudo codes are as shown in the following:

```
Phase 1 Communication via Quantum Channel (Sender)
1:  Generate bit string s randomly (0 or 1)
2:  For each bit s then
3:      choose base randomly resulting base b[i]
4:  End for
5:  For each bit s then
6:      generate a photon
7:      If s[i] = 0 and b[i] = R then p[i] = 0°
8:      If s[i] = 1 and b[i] = R then p[i] = 90°
9:      If s[i] = 0 and b[i] = D then p[i] = 45°
10:     If s[i] = 1 and b[i] = D then p[i] = 135°
11:     send p[i] to Receiver
12: End for
Phase 2 Communication via Quantum Channel (Receiver)
1:  For each photon p'[i] received then
2:      generate RANDOM base result b'[i]
3:      measure photon p'[i] in respect to base b'[i]
4:      result bit s'[i]
5:  End for
Phase 3 Communication via Classical Channel
1:  For each bit s'[i] and s[i] then
2:      If base b'[i] ≠ b[i] then eliminate bit s'[i]
3:      If base b[i] ≠ b'[i] then eliminate bit s[i]
4:  End for
Phase 4 Determine the presence of eavesdropper
1:  For a subset of string bits s
2:      If s'[i]  ≠ s[i] and b'[i] = b[i] then
3:          Eavesdropper detected
4:      End if
5:      discard bit string s[i] and s'[i]
6:  End for
```

3.3 Simulation Tool

To simulate a QKD device, a quantum security system was developed in Java NetBeans IDE to simulate the process of Quantum Information such as quantum polarizer and eavesdropping in the quantum channel. The user is able to select the length of the bit strings and opt to enable eavesdropping in the quantum channel as shown in Fig. 2.

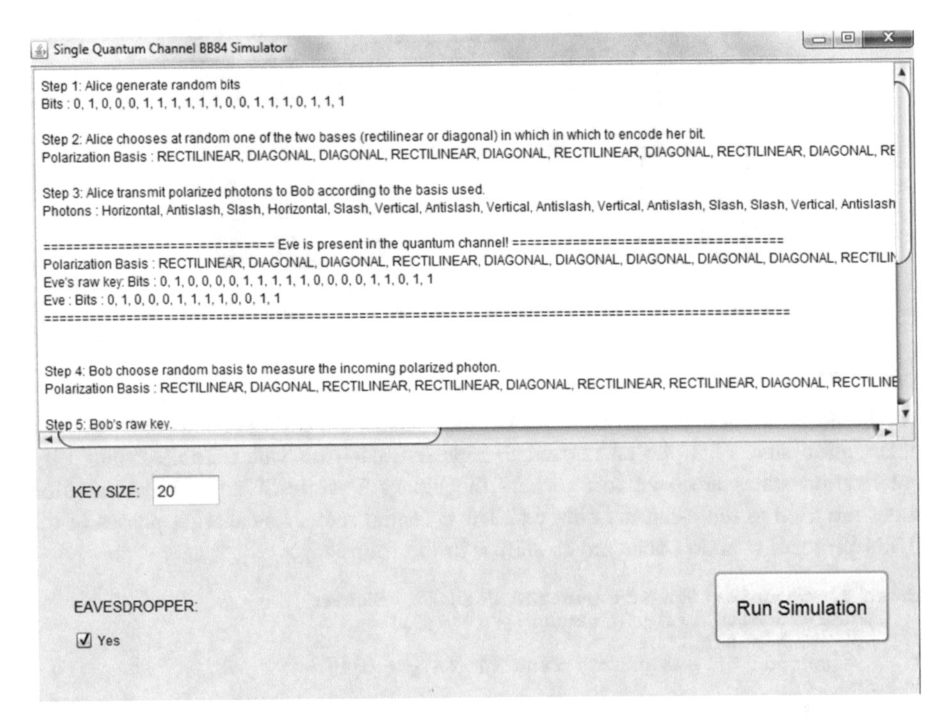

Fig. 2. Single-quantum channel simulation interface

4 Results and Discussions

Simulated data and results on QBER and raw key efficiency based on the single quantum channel key distribution method are presented in this section based on the simulation tool. In this simulation, it is assumed that the user possesses a perfect quantum channel without noises or disturbances and a perfect single-photon source. Non-perfect simulation shall be done in next phase of this research work.

4.1 Simulated QBER and Raw Key Efficiencies for BB84 Protocol

In order to calculate Quantum Bit Error Rate, Eq. 3 was used to estimate the probability of percentage of error across the quantum channel during key distribution. Figure 3 shows 100 simulations on the BB84 protocol to obtain the average QBER based on 200 qubits.

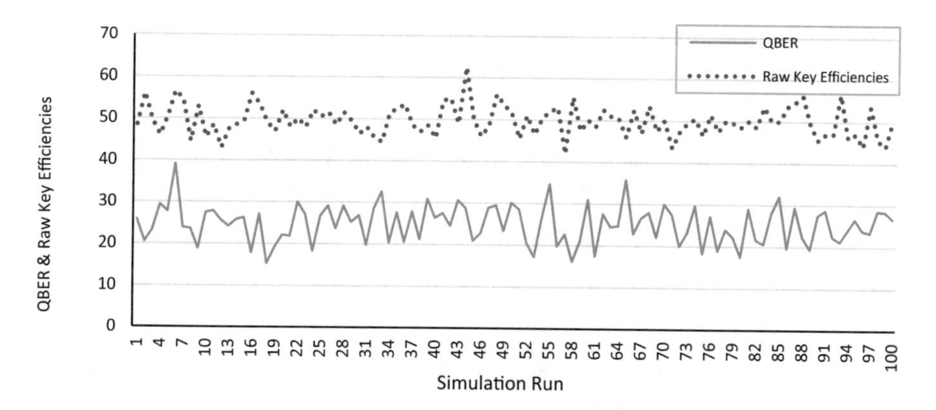

Fig. 3. QBER and raw key efficiency simulation results on BB84 protocol

From 100 simulations, the average QBER can be calculated using the following equation.

$$\bar{x} = \frac{\sum_{i=1}^{n} x_i}{n} \tag{5}$$

where n is the number of simulations and x is the value obtained from each simulation.

The average QBER on BB84 protocol was found to be 25% in the presence of an eavesdropper. In other words, an eavesdropper present in the quantum channel may have 25% knowledge on the transmitted key.

From the simulated outcome, the average QBER of the BB84 protocol is consistent with the findings reported by Nedra Benletaief et al. [21], Mohamed Elbokhari et al. [22] and Di Jin et al. [8].

To determine raw key efficiencies, Eq. 4 was used. 100 simulations on BB84 protocol were conducted to measure the average raw key efficiencies based on 200 qubits. The simulation results are shown in Fig. 3.

The average raw key efficiencies was calculated to be 50% by Eq. 5. This was expected, as at least half of the generated bits will be discarded due to the consequences of not knowing which bases that Alice chooses to encode the photons. Hence, Bob was forced to randomly choose 1 of the 2 bases to measure an incoming photon, resulting in 50% probability of choosing the correct base. The average raw key efficiencies obtained from this simulation was agreeable to the findings reported by Di Jin [8], Chris Erven et al. [23] and Hoi-Kwong Lo et al. [24].

5 Conclusion

In conclusion, the simulation study has reaffirmed that the results of QBER and raw key efficiencies are 25% and 50% respectively for the single quantum channel BB84 protocol. These simulation results were consistent with the findings reported by other researchers.

For future work, dual quantum channel security model will be proposed as a new method to improve Quantum Bit Error Rate (QBER) and raw key efficiencies based on the single-quantum channel communication model.

Acknowledgment. This research project is funded by the Ministry of Education Malaysia (MOE) Fundamental Research Grant Scheme (Grant No: 0153AB-K42).

References

1. Aaronson, S.: The limits of quantum computers. Sci. Am. **298**, 62–69 (2008)
2. Gambetta, J.M., Chow, J.M., Steffen, M.: Building logical qubits in a superconducting quantum computing system. npj Quantum Inf. **3**(2), 1–7 (2017)
3. Barreno, M.A.: The future of cryptography under quantum computers. Dartmouth College Computer Science Technical Reports (2002)
4. Murugesan, S., Colwell, B.: Next-generation computing paradigms. Computer **49**, 14–20 (2016)
5. Bennett, C.H., Brassard, G.: Quantum cryptography: public key distribution and coin toss-ing. Theor. Comput. Sci. **560**, 7–11 (2014)
6. Wiesner, S.: Conjugate coding. ACM SIGACT News **15**, 78–88 (1983)
7. Brassard, G., Lütkenhaus, N., Mor, T., Sanders, B.C.: Limitations on practical quantum cryptography. Phys. Rev. Lett. **85**, 1330 (2000)
8. Jin, D., Verma, P.K., Kartalopoulos, S.V.: Fast convergent key distribution algorithms using a dual quantum channel. Secur. Commun. Netw. **2**, 519–530 (2009)
9. Gao, J., Zhu, C., Xiao, H.: Efficient quantum key distribution scheme with pre-announcing the basis. EPL (Europhys. Lett.) **105**, 60003 (2014)
10. Basu, S., Sengupta, S.: Modified BB84 protocol using CCD technology. J. Quantum Inf. Sci. **6**, 31 (2016)
11. Amiri, P.: Quantum computers. IEEE Potentials **21**, 6–9 (2002)
12. Rodney, V.M., Simon, J.D.: The path to scalable distributed quantum computing. Computer **49**(9), 31–42 (2016)
13. Oraevsky, A.N.: On quantum computers. Quantum Electron. **30**(5), 457–458 (2000)
14. Wootters, W.K., Zurek, W.H.: A single quantum cannot be cloned. Nature **299**, 802–803 (1982)
15. Heisenberg, W.: Physics and Philosophy. Prometheus Books, New York (1999)
16. Wiedemann, D.: Quantum cryptography. ACM SIGACT News **18**, 48–51 (1986)
17. Li, H.-W., Yin, Z.-Q., Wang, S., Qian, Y.-J., Chen, W., Guo, G.-C., Han, Z.-F.: Random-ness determines practical security of BB84 quantum key distribution. Sci. Rep. **5**, 16200 (2015)
18. Mogos, G.: Quantum key distribution protocol with four-state systems-software imple-mentation. Proc. Comput. Sci. **54**, 65–72 (2015)
19. Senekane, M., Mafu, M., Petruccione, F.: Six-state symmetric quantum key distribution protocol. J. Quantum Inf. Sci. **5**, 33 (2015)
20. Mafu, M.: A simple security proof for entanglement-based quantum key distribution. J. Quantum Inf. Sci. **6**, 296 (2016)
21. Benletaief, N., Rezig, H., Bouallegue, A.: Toward efficient quantum key distribution reconciliation. J. Quantum Inf. Sci. **4**, 117 (2014)

22. Elboukhari, M., Azizi, M., Azizi, A.: Quantum key distribution protocols: a survey. Int. J. Univ. Comput. Sci. **1**, 59–67 (2010)
23. Erven, C., Ma, X., Laflamme, R., Weihs, G.: Entangled quantum key distribution with a biased basis choice. New J. Phys. **11**, 045025 (2009)
24. Lo, H.-K., Chau, H.F., Ardehali, M.: Efficient quantum key distribution scheme and a proof of its unconditional security. J. Cryptol. **18**, 133–165 (2005)

Audit Plan for Patch Management of Enterprise Applications

Lois Odilinye, Sergey Butakov(✉), and Shaun Aghili

Concordia University of Edmonton, Edmonton, AB, Canada
clodilin@student.conccordia.ab.ca,
{sergey.butakov, shaun.aghili}@concordia.ab.ca

Abstract. Patch management is a risk management tool for enterprises and a key element of IT security programs. Improper patch management may lead to downtime and interruption, data leakage, penalties and fines for noncompliance with regulations, lost revenue, damaged reputation, litigation fees, etc. It is imperative for enterprises to develop, implement and monitor a well-structured patch management program. Monitoring of program implementation includes its audits. In this paper, an audit program and plan for patch management of enterprise applications is developed. Program includes common elements recommended by information security frameworks and the research community. The audit plan includes audit areas, accompanying audit objectives and tests. Finally, the audit areas, audit objectives and audit tests is mapped to applicable sections of the NIST cybersecurity framework.

Keywords: Patch management · Audit objective · Audit tests · Risk assessment

1 Introduction

1.1 Background

Modern business environment landscape is witnessing ever increasing complexities as more and more businesses largely depend on information systems. Enterprise applications and the associated technologies are relied on to support business processes and operations. As a result, the security of enterprise applications has become critical and attracts great attention from businesses and the research community.

According to [1] an information systems journal, enterprise applications are crucial for the successful and effective operations of business activities as it provides complete and timely information for management decisions. Enterprise information systems ensure effective data exchange between business partners and ultimately enhance efficiency and productivity through business level support functionality. Unfortunately, the growing number of attacks on such systems negatively affects the benefits of IT.

K.J. Kim et al. (eds.), *IT Convergence and Security 2017*,
Lecture Notes in Electrical Engineering 450,
DOI 10.1007/978-981-10-6454-8_22

Worms and malicious code among other factors that exploit known vulnerabilities of unpatched applications and system software is growing at an unprecedented pace. Therefore, enterprises have to put in place a sound patch management program to mitigate the risks associated with vulnerabilities and ultimately taking proactive steps to maintain the success of its patch management efforts. For example, a crucial element of maintaining patch management efforts is continuous monitoring, which typically includes audit and assessment.

2 Patch Management and Associated Risks

In the last decade, where high profile, high-cost security incidents are prevalent adequate defense-in depth mechanisms to effectively secure applications is of utmost importance. One of the essential elements of an information security program is patch and vulnerability management. The judgement call for proper patching came with massive outbreaks of #Wannacry and #Petya ransomware in 2017 that affected hundreds of thousands of systems across the globe leading to massive data losses in healthcare, government, banking, etc. in hyper-connected societies. In both outbreaks, the malware exploited vulnerability for which the patch was released few weeks prior to attack.

According to [2] patch management processes and procedures is a core security layer in any organization's defense-in depth strategy. Patch deployment is a subset of the software maintenance phase in the software development lifecycle. Software lifecycle typically involves planning, systems analysis, conceptual systems design, system evaluation and selection, detailed design, application programming and testing, software implementation/ deployment and software maintenance [3]. Patch management is the phase in software development lifecycle that ensures proper software maintenance is provided.

Patch management is defined as a process of identifying, acquiring, installing and verifying patches for software and systems [4]. Usually patches does one of the following, fix a bug, install new drivers or other supplementary software, address new security vulnerabilities, address software stability issues, and upgrade the software.

2.1 Risks Associated with Improper Patch Management Programs

A good patch management program entails developing effective plans and policies for patching which covers what patches should be applied to which systems and at what time. [5] Outlines the following as risks of poor patch management program: delayed or non-applied patches (SQLSlammer, #Wannacry and #Petya), invalidated patches, insufficient testing, downtime and interruption, vulnerabilities in patch management tools, no fallback procedures, improper patch identification and installation.

Literature outlines common reasons for patch management program failures: lack of corporate policy requiring patching; vague understanding of roles and responsibilities associated with patching; wrong expectation of scope; poor software lifecycle management; no release or change management maturity; one patch solution for all needs; lack of tools or automation to support automation in a repeatable manner, no computer build standard or accreditation for new computers [6].

The risks associated with a poor patch management program ultimately leads to negative implications for organizations such as data loss, fines for non-compliance with regulatory requirements, damaged reputation, stolen intellectual property and litigation fees.

2.2 The Need for Patch Management Audit

A review of the risks associated with poor patch management makes it obvious that it is essential for organizations to have a system that assesses and monitors the extent and success of patch management efforts. Auditing the patch management program enables evaluation and assessment of the specific aspects of the patch management program. Thus, auditing provides an efficient tool to measure deviations from patching targets and objectives. Patch management audit objectives and tests should be aimed at demonstrating that an enterprise patch management efforts are effective and efficient. This enables meaningful assessment of the security posture of an enterprise with regards to patching [2].

The recommendations reviewed recommended the following elements of a sound patch management program: inventory of information assets, patch management policy, optimal timing of patches, risk assessment, identification of new vulnerabilities and patches, testing and deploying patches.

A few recommendations included different considerations or elements of a patch management process. In [7] authors recommended that a configuration management plan, disaster recovery plan, and incidence response plan should be included in the patch management program. Authors of [6] and [8] suggested the creation of a patch and vulnerability group to monitor new vulnerabilities and patches. Work [9] suggested a flaw remediation process incorporated into configuration management process. This process would identify, correct and report information system flaws, establish benchmarks for taking corrective actions amongst other functions [9].

While various recommendations have been put forward by information systems security frameworks and the research community, there is no existing attempts for an overarching template that would cover all major areas in patch management. This research developed such an audit plan template focused on patch management program. The proposed audit plan has key audit areas, audit objectives and potential audit tests to enable IT auditors evaluate the patch management program and effectively measure deviations from patch management objectives.

3 Summary of the Key Elements of the Audit Plan

Key elements of the audit plan include the following:

Risk Assessment: Timely response is critical to successful patch management efforts. As a result of limited resources available to the enterprise, it is important to determine which systems should be patched first. Conducting a risk assessment helps to prioritize patching. According to [10] risk assessment for patch management should be done in the following areas:

Threat level: here the severity of the threat is considered, the likelihood of it affecting the business environment.

Vulnerability: this area considers the level of vulnerability of applications and systems. Application and systems that are more vulnerable may include systems that outside the perimeter firewalls.

Criticality: this area considers how important a system is to the business environment. Mission critical applications and systems is a key risk area. Examples include mail servers, database servers among others.

Thus, mission-critical systems, vulnerable systems and severity of the threat should be of utmost priority.

3.1 Applicable Categories and Sub-categories of the NIST Cybersecurity Framework

The key audit areas and the accompanying audit objectives and audit tests have been mapped to relevant categories and sub-categories in the NIST cybersecurity framework that relate to patch management program. The categories include governance, risk assessment, asset management, information protection processes and procedures, data security, and protective technology.

As shown in Table 1, patch management policy is mapped to the governance category (Governance ID-GV: The policies, procedures and processes to manage and monitor the organization's regulatory, legal, risk and operational requirements are understood and inform the management of cybersecurity risks) and sub- categories- ID. GV-1: Organizational security policy is established. ID-GV-2: Information security roles and responsibilities are aligned internal roles and external partners.

Table 1 represents the actual template – major contribution of this research project. Table elements can be interpreted as described below in this paragraph. For example, the first area is policy. It is related to ID-GV domain in NIST framework [14] thus can be included in the Governance auditing as well. To verify the existence of the policy the auditor would need to look at the major required elements of the policy as outlined in Audit Tests column: purpose, scope, roles and responsibilities, management commitment, and compliance should appear in the policy. Next area is connected to risk assessment domain in NIST framework (ID.RA) thus can be included into the auditing procedures for Risk Management. As in the previous domain, tests are also provided in the table. As it can be seen from this short description condensed format provided in Table 1 can be easily converted to an extended checklist that an auditor can run through.

Table 1. Proposed audit plan alignment to NIST cybersecurity framework

Key audit Areas	Audit objectives	Audit tests/Procedures	Mapping to applicable NIST CSF categories
Patch management policy	Existence of a patch management policy that addresses; purpose, scope, management commitment, and compliance	Obtain and examine current policy document to ensure purpose, scope, roles and responsibilities, management commitment, compliance are addressed	**Governance (ID-GV):** The policies, procedures and processes to manage and monitor the organization's regulatory, legal, risk and operational requirements are understood and inform the management of cybersecurity risks **ID.GV-1:** Organizational security policy is established. **ID-GV-2:** Information security roles and responsibilities are aligned internal roles and external partners
	Defined scope for patch management	Examine policy document to ensure that systems and applications that need to be patched are clearly defined	
	Adequate roles and responsibilities for patch management	Review policy document to ensure roles and responsibilities are clearly defined. Interview information security personnel charged with patching roles and responsibilities to confirm adequacy of roles and responsibilities	
	Frequency of review and updates of patch management policy	Obtain and examine previous policy documents to determine if organization-defined frequency of reviews and updates have been adhered to	
Risk assessment	Risk assessment of new patches to determine severity levels and impact on information assets	Obtain and examine risk assessment reports to verify if risk assessments for patches are conducted based on threat level, vulnerability and criticality of security patches	**Risk Assessment (ID.RA):** The organization understand the cybersecurity risks to organizational operations (including mission, function, image, and reputation), organizational assets and individuals. **ID-RA-1**: Asset vulnerabilities are identified and documented. **ID-RA-2:** Threats and vulnerabilities is received form information sharing forums and sources. **ID-RA-3:** Potential business impact and likelihood are identified and documented
	Effective notification tracking system for release of new patches	Examine notification tracking systems: subscription to vendor's mailing list, security bulletins on software enhancement and security	
	Identification of potential impacts based on the risk assessment of patches	Examine current and previous business impact analysis reports to confirm risk levels and impact on organization's systems and applications	
	Conduct periodic vulnerability assessments	Verify that scanning tools have successfully completed their weekly or daily scans for the previous 30 cycles of scanning by reviewing archived alerts and reports to ensure that the scan was completed	

(*continued*)

Table 1. (*continued*)

Key audit Areas	Audit objectives	Audit tests/Procedures	Mapping to applicable NIST CSF categories
	Comprehensive and up-to-date inventory of information assets	Review previous assets inventory database reports to confirm frequency of updates and completeness of the report	**Asset Management (ID. AM):** The data, personnel, devices, systems and facilities that enables the organization to achieve business purposes are identified and managed. **ID-AM-2:** Software platforms and application within the organization are inventoried. **Information Protection Processes and Procedures (PR. IP):** Security policies, processes and procedures are maintained to **PR.IP-3:** Configuration change control processes are in place. **PR.IP-4:** Backups of information are conducted, maintained and tested periodically. **Data Security:** Information and records (data) are consistent with the organization's risk strategy to protect the confidentiality, integrity and availability of information. **PR.DS-6:** The development and testing environment(s) are separate from the production environment
	Change control process for patches	Review change request forms for previous patches to ensure adequate approval and documentation by appropriate IT personnel	
	Successful backups of mission-critical applications and systems before patch deployment	Select at least five systems within the enterprise network environment and restore to a test system using the most recent backup. Verify that the system has been restored properly by comparing the restore results to the original system	
	Patch testing and deployment	Review test plan/guidelines and test results to verify the success of patch testing. Post-deployment reports should be reviewed to ensure that there are no reported issues at least one week after patch application	
Examination of security controls for exception	Existence and effectiveness of network protection systems for applications and systems where patches cannot be installed or vulnerabilities cannot be resolved [14]	Take a software test program that appears to be a malware, but not included in the authorized software list, to at least 10 randomly selected systems within the network. Verify that the systems generate an alert or e-mail notice regarding the malware within one hour. Verify that the system generates an alert or e-mail indicating that the malware is blocked or quarantined [16]	**Protective Technology (PR. PT):** Technical security solutions are managed to ensure the security and resilience of systems and assets, consistent with related policies, procedures and agreement. **PR.PT-1:** Audit/logs are determined, documented, implemented and reviewed in accordance with policy

4 Conclusion

Patch management program is an essential requirement for a robust and successful information security program as it plays critical role in protecting confidentiality, integrity and availability of any enterprise's information systems.

Thus, the research proposes an audit plan for patch management, which included the following major audit areas: first, patch management policy, which includes the scope of systems and applications to be patched, roles and responsibility amongst others. Second, risk assessment, typically involving risk prioritization/classification and impact analysis with regards to patching. Third, key patch management processes and procedures such as change management process, patch testing and deployment, inventory of information assets. Lastly, examination of security controls, such as network protection systems such firewalls, intrusion detection and prevention systems. Hence, the audit plan provides a tool for IT auditors to consistently assess the success and extent of patch management efforts. Patch management program should not be thought of as a one-time procedure but as a process that requires continuous review and monitoring.

This research focused on the development of an audit plan for patch management and concentrated on major/key audit areas. Future research can be done on the applicability of the audit objectives and audit tests, such as assessment of patching times, scope, and risk assessment in a typical enterprise. This would further validate its suitability of application.

References

1. Serova, E.: Enterprise information system of new generation. Electron. J. Inf. Syst. Eval. **15** (1), 116–126 (2012)
2. MacLeod, K.J.: Patch Management and the Need for Metrics, SANS Institute Reading Room (2004)
3. Hall, J.: Information Technology Auditing and Assurance. South - Western Cengage Learning, Ohio (2011)
4. Souppaya, M., Scarfone, K.: Guide to Enterprise Patch Management Technologies. NIST, Virginia (2013)
5. Joint Universities Computing Centre: Network Computing, Information Security Newsletter, p. 1, 23 December 2013
6. Hoehl, M.: Framework for Building a Comprehensive Enterprise Security Patch Management Program, SANS Institute Reading Room (2013)
7. Tom, S., Christiansen, D., Berrett, D.: Recommended Practice of Patch Management of Control Systems. Idaho National Library, Idaho (2008)
8. Mell, P., Tracy, M.C.: Procedures for Handling Security Patches. National Institute of Standards and Technology, Washington DC (2002)
9. Blank, R.M., Gallagher, P.D.: Security and Privacy Controls for Federal Information Systems and Organizations, National Institute of Standards and Technology (2013)
10. Mell, P., Bergeron, T., Henning, D.: Creating a Patch and Vulnerability Management Program, National Institute of Standards and Technology (2005)

11. Medzich, M.: Deploying a Process for Patch Management in relation to Risk Management, SANS Institute (2004)
12. Ruppert, B.: Patch Management, SANS Institute Reading Room (2007)
13. Liu, S., Kuhn, R., Hart, R.: Surviving Insecure IT: Effective Patch Management. IT Prof. **11** (2), 49–51 (2009)
14. National Institute of Standards and Technology: Framework for Improving Critical Infrastructure Cybersecurity, NIST (2014)
15. Council on Cybersecurity: The Critical Security Controls for Effective Cyber Defense, SANS Institute
16. National Institute of Standards and Technology: Assessing Security and Privacy Controls in Federal Information Systems and Organizations, NIST (2014)

Evading Tainting Analysis of DroidBox by Using Image Difference Between Screen Capture Images

Dae-Boo Jeong and Man-Hee Lee(✉)

Department of Computer Engineering, Hannam University, Daejeon, Korea
jungdaeboo@gmail.com, manheelee@hnu.kr

Abstract. Protecting personal and business data stored in smart phones from information leaking applications becomes very important. To detect such apps as early as possible, the data tracking functionality, called tainting analysis, is being utilized in many areas, and DroidBox with TaintDroid is one of the most frequently used dynamic analysis tools for Android system emulation. In this study, we showed a simple steganographic technique utilizing two consecutive screen captures so that TaintDroid or smartphone users cannot track or detect.

Keywords: Android · DroidBox bypass · Bypassing taint check · Data leak · Steganography · Screen capture · Image difference

1 Introduction

As smart phones are used for both work and personal life extensively, the number of information stealing apps keeps increasing. Different from other malicious behavior, detection of information leakage is very difficult because real computer architecture does not provide details about how data is processed in registers and the main memory.

Tainting analysis in virtual environments can be a good alternative solution for analyzing malwares before they steal data from real companies and users. The technique is to put a tag on data, keep track of the tagged data throughout the execution, and finally report whether the data is leaked or not. TaintDroid is almost the first practical tainting analysis tools for Android [1]. It comes with DroidBox, an Android virtual environment, that is a famous dynamic analysis tool [2,5].

However, TaintDroid cannot protect every data in the system. As TaintDroid developers noted [1], TaintDroid was not designed to track all possible side channel attacks [6,8,9]. G. Sarwar et al. presented many side channel attack methods to evade TaintDroid's tainting analysis [3]. In our previous study [7], we selected and implemented a bitmap attack. Its attack scenario is simple. The first step is to write data of interest on the screen, and then take a screen shot. Since TaintDroid is unable to track data stored on the bitmap memory for the screen, screen

K.J. Kim et al. (eds.), *IT Convergence and Security 2017*,
Lecture Notes in Electrical Engineering 450,
DOI 10.1007/978-981-10-6454-8_23

captures of data written on the screen is a nice method to leak private data. The last step is to send the images to an outside server where text information is extracted by using OCR (Optical Character Recognition).

Even though the use of screen capture could evade TaintDroid successfully, it is not completely undetectable by ordinary smartphone users. That is because it takes at least several seconds to take a screen shot. This means that private data should be displayed for the time. When a user keeps looking at the screen, it is very hard to miss this behavior because the private data appears on the screen suddenly and disappears after some seconds.

In this research, we presented a simple data steal technique better than the previous idea while successfully deceiving both TaintDroid and users. Before explaining how to do it, we need to explain a key finding on which our idea is based. A screen shot image looks similar to its original image, but they are very different when we compare pixel information. By the way, while testing many images on various devices, we found a very consistent symptom: Differences between two images are still conveyed to the two screen captured images. For example, if two images are exactly same except the first pixel's blue value by one, their screen captured images show the same difference on the first pixel with the same blue value.

This study utilized this finding to devise a new data leaking technique as follows: First, we send out a clean picture's screen shot to an outside server. Second, we embed private information into the picture using a simple steganographic technique such that it is impossible to differentiate two pictures by human eyes. Third step is to take a screen shot for the second picture. Fourth, we send out the new screen shot to the outside server where the original private data is recovered by comparing two screen shots. Since our method uses the screen bitmap, TaintDroid still cannot detect it.

The rest of the paper is organized as follows. Section 2 describes a side channel attack, one of the techniques for bypassing tainting analysis in TaintDroid. Section 3 proposes an data leaking algorithm using consecutive screen capture based on side channel attack techniques. Sections 4 and 5 develop a data leaking algorithm and validate the data leaking algorithm through testing. Finally, Sect. 6 presents our conclusions.

2 Related Work

Sarwar et al. evaluated how effective TaintDroid tracks information [3]. As noted in TaintDroid authors, TaintDroid does not implement control dependence information tracking. In this method, tainted data itself is not propagated. Rather, clean data is assigned to new variables, but the decision of clean data depends on tainted data, thus indirectly copying tainted data to untainted variables. Second approach is to use benign code trusted by system. The benign codes are not perfectly tracked, so various activities of the code are not tracked. Third approach is side channel attack. Side channel attack in cryptography is to extract physical implementation through timing information, power consumption, or sound rather than through brute force or theoretical attack on algorithm and its cryptographic system.

Side channel attack in TaintDroid is any techniques to gain access to private data not through normal data access method, but through other method out of scope of TaintDroid. Common timing attack is possible in TaintDroid by running a special task with some meaningful intervals. By monitoring the task's execution intervals, any private data can be leaked without detection. Final approach is to use bitmap cache that is used for screen display. Since TaintDroid stops tracking on bitmap cache, any data written on screen can be exposed to outside without detection. In our previous research [7], we adopted this idea and show how font sizes and types affect the effectiveness of data leakage.

Main disadvantage of our previous idea is that private information that we want to steal shows on the screen so that bitmap cache can recorded via screen capture. If a user looks at the screen carefully, he can recognize that some texts appear for seconds and disappear suddenly.

3 Data Leak by Using Consecutive Screen Capture

The key problem of using the screen capture is that data of interest needs to be displayed to the smartphone user, inevitably making the user suspicious about possible maliciousness of the app. This might be the most unwanted situation by the hacker.

To solve this problem, we propose to use consecutive screen captures, which is based on an interesting symptom of screen capturing behavior of many smartphones: differences between two images are exactly same as the differences between two screen captured images.

Following is how to utilize the symptom to leak data without detection by TaintDroid or human users. Its overall schematic is depicted in Fig. 1. First, we display a normal image called a cover image without any private information on the screen and capture it. Second, we generate a stego image by embedding private information into the original image. Then, the stego image is shown to the user and captured. In this process, it is important not to embed too much information. In the field of image processing, provided the bit depth is 8 bits, the PSNR (Peak Signal-to-Noise Ratio) of the original and its compressed or lossy image needs to be more than 30 dB in order that human eyes are unaware of the difference between the two images. Then, we send out the two captured images to the predestined server where differences of RGB values of the images are extracted to recover the private information.

Please note that this skill looks similar to steganographic techniques to conceal private information in digital images [3, 4, 10, 11]. The main difference is that we use screen captured images rather than normal images. In the digital image steganography, pixel values at specific locations are modified and the image is sent out from the system. The image recipient can extract hidden information because the image was directly transferred. However, in our situation, the image cannot be sent out the Internet directly since all the data was tracked.

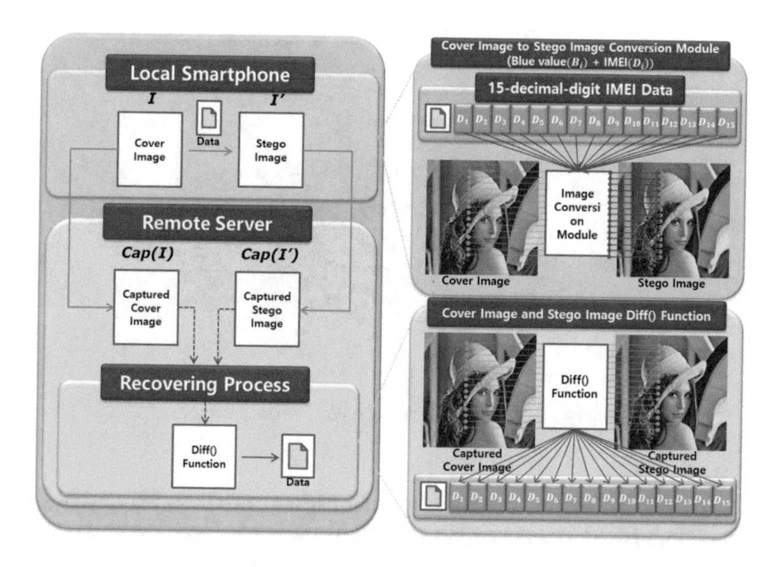

Fig. 1. Data leak concept using two captured images

It would be best if an original and its captured images are perfectly the same, but that is not the case; the original and its captured images are not same at all. In Android emulator, we captured a Lenna image and compared it with the original image shown in Fig. 2. They look the same to human eyes since the PSNR of two images is about 45.52 dB, but the graph in Fig. 3 shows difference between two images. The 145×145 image has 21,025 pixels and each pixel has red, green, and blue values from 0 to 255. We compared RGB values of all pixels of two images and calculated RGB color difference by adding up absolute difference value of each color. For example, when RGB values of the two pixels of the same location from the two images have (128, 128, 128) and (129, 127, 128) respectively, the pixel's RGB color difference is two. 76.23% of pixels have different RGB values and the pixel's average color difference is 3.61 with a standard deviation of 3.27. Therefore, it seems impossible to recover the hidden information from captured image because the information loss occurs when capturing the screen image.

Fig. 2. Original (left) and captured (right)

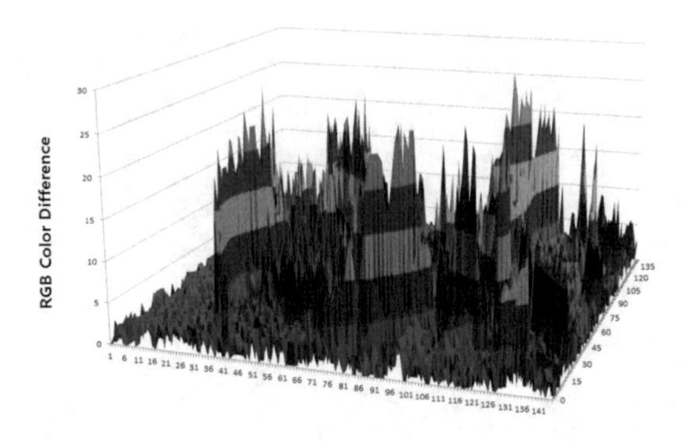

Fig. 3. Color difference of Lenna images

As mentioned earlier, we use two captured images to leak private information. For this method to be practical, it should be always true that we can recover from the two captured images the exactly same information that we embed into original images. For this, we need to assert that the following formula, F, should be true all the time. Its main concept is depicted in Fig. 1.

$$F(I,\, I') : Diff(I,\, I') \equiv Diff(Cap(I), Cap(I'))$$

where I is an original image, called cover image, I' is a stego image of I, to which data was inserted, *Cap(I)* is a captured image of I, *Diff()* function produces the pixel difference information of two input images, I and I'. The information includes every pixel's RGB difference as well as the percentage of different pixels, average color difference, and standard deviation. $\equiv$ is defined that two *Diff()* functions produce the same results, meaning that differences between two images are exactly same as the differences between two screen captured images, asserting *F(I, I')* is true. After all, we make sure that the inserted information in I' can be recovered from the difference between *Cap(I)* and *Cap(I')*.

4 Validation by Experiments

To verify whether the formula is true in other environments, we tested six randomly selected images shown in Fig. 4 on three different real devices: LG G2 (LG-F320K), Galaxy Nexus (SHW-M420S), Galaxy S2 (SHW-M250S). For each image, we randomly chose one pixel and changed its blue value by one. Then, we captured two images to calculate *Diff(I, Cap(I))* and *F(I, I')* on three devices.

Table 1 shows the test results. In Lenna image captured in LG G2 smart phone, its *Diff* percent is 77.23%, meaning that percent of pixels of caputred image are different from the original pixel values. Its average color difference is 3.46. With the same Lenna image, Galaxy Nexus and S2 showed different *Diff* percent and average color difference. In all other tests, *Diff* percent and average color difference look random. However, *F(I, I')* is true all the time, meaning that

(a) Lenna (b) Big Boss (c) Carolina (d) Scarlet (e) Veteran (f) Shin chan

Fig. 4. Six randomly selected images used for testing

Table 1. Images from the real devices test results

Image name	LG G2			Galaxy Nexus			Galaxy S2		
	Diff percent	Avg color difference	F(I, I')	Diff percent	Avg color difference	F(I, I')	Diff percent	Avg color difference	F(I, I')
Lenna	77.32%	3.46	True	97.54%	8.14	True	97.56%	7.99	True
Bigboss	49.13%	3.83	True	84.80%	7.94	True	83.24%	7.68	True
North Carolina	38.63%	6.52	True	56.90%	10.94	True	56.56%	10.84	True
Scarlet	66.68%	2.87	True	93.94%	5.68	True	94.63%	5.77	True
Veteran	28.66%	2.54	True	61.39%	3.34	True	58.26%	3.42	True
Shin chan	61.94%	4.27	True	96.88%	12.41	True	94.68%	12.31	True
Average	53.73%	3.91	All true	81.91%	8.08	All true	80.82%	8.00	All true

the pixel difference information between two original images can be recovered from the captured images.

To support our idea, we collected 100 images randomly from Google and performed the same tested in the Android emulator. Its average *Diff* percent is 49.94% with standard deviation of 19.64. Its average color difference is 4.59 with standard deviation of 2.03. Still, *F(I, I')* stays true in all cases.

With this results, we reasonably assume that capture algorithms implemented in each smart phone are different, but their algorithmic behavior is so static that pixel difference information is conveyed in multiple captured images.

5 Implementation of Data Leakage App Using Two Captured Images

From now on, we are ready to devise a data leaking algorithm based on what we just found. Here, just for a proof of concept, we developed a simple IMEI (International Mobile Equipment Identity) leakage technique and showed how it worked.

We used the Lenna image. Since IMEI is a 15-decimal-digit number, we randomly selected 15 points and changed its blue value by each digit's decimal value, which is similar to the traditional steganographic algorithms changing least significant bits of pixel valuses. Since digits can be zero, we added one to the digits. The reason why we chose pixel locations randomly is that it would decrease the possibility of being detected.

Once the modification is done, the next step is to show the original Lenna image to the screen and capture it, then do the same process with the modified image. Now two captured images are available. We sent them out to a remote server where the leaking information is recovered by comparing two captured images. This process is shown at the right side of Fig. 1.

Please note that it is not necessary to send the two captured image to the outside. That is, the difference information can be recovered in the local smart phone and the extracted information can be directly sent to the outside. However, packets in transmission can be captured or monitored by data loss prevention and detection systems. In this situation, private information leakage can be detected. To avoid this situation, the information can be encrypted beforehand, but the use of cryptographic APIs or sending out arbitrary strings with high entropy values could attract unnecessary attentions. Therefore, sending out two similar images that look the same would be better for lowering the possibility of being detected.

We developed a sample app and a server for this test. After displaying the Lenna image on the screen, the app captures a screen image and show the next image after changing 15 locations' pixel values according to IMEI value. After capturing once again, the app sends the two captured images to the outside server. We successfully recovered the inserted IMEI. Figure 5 shows a Json file generated by DroidBox for application analysis. *opennet* part reports that the application made two connections with 203.247.39.97, our test server, and sent two files via the port number 22, but *dataleaks* part reports nothing, which means that any data leak events did not occur. With this results, we showed that we successfully leaked the IMEI by using two captured images while TaintDroid and users cannot detect the event.

We iterated the same test for the one hundreds images that we used for *Diff()* function validation. Average PSNR values between the cover and stego images is about 72.95 with a standard deviation of 0.49. Therefore, we can say that this process is so quick and the images are so undistinguishable that users cannot notice that the image has been changed.

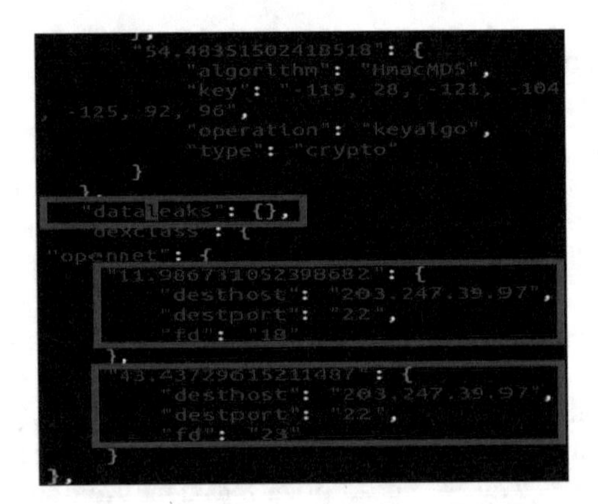

Fig. 5. DroidBox detection results

6 Conclusion

The more people keep business and private data stored in smart phones, the more important it becomes to protect the data from theft. DroidBox with TaintDroid is a well-known Android virtual environment with strong tainting analysis functionalities. Bitmap memory for screen is a known security hole that TaintDroid cannot keep tracking tagged information well. In our previous research, we proposed a method to leak private information by capturing the screen with private information displayed in form of text and applying OCR to extract the text from the captured image.

In this research, we improved the previous idea's weak point that the private data should be displayed for some seconds so that it might increase the possibility of being detected by users. Instead, we proposed a leaking method by embedding private information into the image and capturing the original and data-embedded images separately. We found that the difference information extracted from the two captured images is exactly the same with that from the original and data-embedded images. By using this finding, we successfully developed a private data leaking application evading TaintDroid's data leak detection mechanism.

We are currently investigating how to detect our data leaking technique proposed in this paper. We also keep searching data leak holes that TaintDroid cannot detect. Through this research, we hope to help build more complete TaintDroid.

References

1. Enck, W., Gilbert, P., Chun, B.-G., Cox, L.P., Jung, J., McDaniel, P., Sheth, A.N.: TaintDroid: an informatin-flow tracking system for realtime privacy monitoring on smartphones. OSDI **10**, 255–270 (2010)
2. Google code - DroidBox. https://code.google.com/p/droidbox/
3. Sarwar, G., Mehani, O., Boreli, R., Ali Kaafar, M.: On the effectiveness of dynamic taint analysis for protecting against private information leaks on android-based devices. NICTA Technical report RT-7091 (2013)
4. wikipedia – side channel attack. https://en.wikipedia.org/wiki/Side-channel_attack
5. Lantz, P.: Droidbox - android application sandbox, February 2011
6. Shuba, A., Le, A., Gjoka, M., Varmarken, J., Langhoff, S., Markopoulou, A.: Demo: AntMonitor: a system for mobile traffic monitoring and real-time prevention of privacy leaks. In: Proceedings of the 21st Annual International Conference on Mobile Computing and Networking, Paris, France, 7–11 September 2015
7. Kim, Y.-K., Yoon, H.-J., Lee, M.-H.: Stealthy information leakage from android smartphone through screenshot and OCR. In: International Conference on Chemical, Material and Food Engineering, August 2015
8. Song, Y., Hengartner, U.: PrivacyGuard: a VPN-based platform to detect information leakage on android devices. In: Proceedings of the 5th Annual ACM CCS Workshop on Security and Privacy in Smartphones and Mobile Devices, Denver, Colorado, USA, 12 October 2015

9. Canfora, G., Medvet, E., Mercaldo, F., Visaggio, C.A.: Acquiring and analyzing app metrics for effective mobile malware detection. In: Proceedings of the 2016 ACM on International Workshop on Security And Privacy Analytics, New Orleans, Louisiana, USA, 11 March 2016
10. Cavallaro, L., Saxena, P., Sekar, R.: AntiTaint-analysis: practical evasion techniques against information flow based malware defense. Stony Brook Computer Sclence Dept, November 2007
11. Cavallaro, L., Saxena, P., Sekar, R.: On the limits of information flow techniques for malware analysis and containment. In: Lecture Notes in Computer Science, July 2008

Social Media Information Security Threats: Anthropomorphic Emoji Analysis on Social Engineering

Kennedy Njenga(✉)

Department of Applied Information Systems, University of Johannesburg, Johannesburg, South Africa
knjenga@uj.ac.za

Abstract. The evolution of anthropomorphism and affective design principles in Social Medial has allowed friends and colleagues to create content across organizational settings that now provision for emotions through the popular use of emoji. Of concern to information systems security practitioners is that the use emoji can be effective in social engineering through facilitating escalated malevolent attacks to unsuspecting victims. The article applies theories from social psychology, criminology and information systems while using the Elaboration Likelihood Model (ELM) to determine the possibility of using emoji as tools for effective social engineering. A university setting was used and student–actors enlisted to execute social engineering scenarios under carefully controlled environments. Screen shots of social engineering using emoji were taken by student-actors and sent to researcher for analysis. Qualitative data analysis involved prepossessing emoji data through tokenization and normalization. Results reveal two important findings. Firstly, that the effective use of emoji is more likely to persuade victims because of unsuspecting emotional appeal. Secondly, more time was taken to persuade a victim when only textual words instead of emoji were used in the interaction process. The results of findings are discussed in the main article.

Keywords: Social media · Social engineering · Emoji · Information security

1 Introduction

1.1 Evolving Social Media

The number of advanced Social Media (SM) software technology applications such as *WhatsApp*™, *Facebook*™ and *Twitter*™ available on the Internet that promote user generated content, has made SM evolve from being a purely social platform to that which is integral to organizational settings [1]. The evolution of anthropomorphism and affective design principles in SM content (such as *emoji*) has allowed friends and colleagues to create SM content across organizational settings that now provision for sentiments and emotions. Anthropomorphic and affective designs in SM content generation has been one of the best successful ways that has inspired emotional elicitation. The use of e*moji* is one of the best examples of successes in SM Anthropomorphic

K.J. Kim et al. (eds.), *IT Convergence and Security 2017*,
Lecture Notes in Electrical Engineering 450,
DOI 10.1007/978-981-10-6454-8_24

designs. *Emoji* were created in the 1990s by NTT DoCoMo, the Japanese communication firm to complement missing face to face and voice tone interactions [2]. An information security issue arises when unauthorized third party users manipulates emotions to gain access to private information (social engineering) [3]. Social engineers will appeal to emotions when malevolently interacting with potential victims in order to escalate these emotions towards malevolent intentions (e.g. fraud, espionage and damage to assets) [4].

Social Media and Behavioral Information Security. Research into behavioral information security often involve determining causes for security incidents such as neutralization [5] and rationalization [6]. There has also been research that focusses intervention and raising security awareness [7, 8] and those that focus on punishment and reward [9]. However, such research has not addressed security in context to social engineering and social media emotional perspectives. Social engineers have used emotions to persuade victims to provide sensitive information and it would be important to establish whether these same techniques can be heightened when anthropomorphic emoji are exploited and used as conduits for social engineering. The following is therefore a research question that could be explored. *Would anthropomorphic emoji posted via social media platforms construe as exploitable information security threats by social engineers?*

Using frameworks from social psychology literature regarding *emotions* and *trust*, information systems security and criminology literature regarding *social engineering*, the work devices a model to explain the likelihood of *emoji* being used by social engineers to exploit victims.

2 Social Media Anthropomorphism and Affective Design

Anthropomorphic analysis would involve computationally classifying/categorizing qualitative emoji data and eliciting emotional attitudes that could be positive, negative, or neutral.

2.1 Social Media Anthropomorphic Analysis

SM users are gradually mastering emergent technology interfaces that offer new possibilities for interaction and content creation. An interesting and emerging area of content generation is the role of user emotions (*emoji*) within the interaction process. When interface design assigns human traits to inanimate objects, this is classified as anthropomorphism [10]. The application of anthropomorphism to SM and applications has increasingly lead to augmented user interaction across various in communities.

Fig. 1. Emoji conveying anthropomorphism and human-like emotional qualities [11]

3 Social Engineering

3.1 Anthropomorphism Engineering

Social engineers often attempt to convince prospective victims by appealing to strong emotions such as excitement or fear [4]. Social engineering is the art and science of getting people to comply to wishes [12, 13]. Compliance is achieved by using "*psychological tricks in order to obtain information needed to escalate access to more information*" [14]. Social engineers will repeatedly take advantage of established interpersonal relationships by creating, influencing and shaping an environment of perceived trust and commitment [15]. Some people will willingly provide sensitive information despite being aware that it could be a security threat [16]. This can be explained when anthropomorphic *emoji* data is subjected to deeper analysis.

3.2 Emoji Analysis

Eliciting Emotions from Anthropomorphic Data. The process of understanding and analyzing anthropomorphic data would involve examining emotional affective representations *(emoji)* and extracting actionable patters and trends from raw social media data. Performing an anthropomorphic emoji analysis on a platform such as Facebook would not differ much from the typical sentimental analysis commonly used in classifying public posts such as *Twitter Sentimental Analysis tool* (DatumBox API). The process would for instance involve creating an API which would make it possible to fetch public anthropomorphic emoji posts that would be filtered and evaluated for sentiment polarity of emoji.

Emoji Consistency. Anthropomorphic emoji analysis would involve determining sentiment polarity. Emoji data is highly correlated with sentiment polarity of posts and words. *Emotional Consistency Theory* (ECT) suggests that the use of two frequently co-occurring words and emoji (emoji and emoji and/or words and words) should have similar sentimental polarity. Emoji sentiment score **D** may be determined as;

$$\mathrm{Dij} = ||\mathrm{ei} - \mathrm{ej}||^2$$

where **ei** and **ej** represent sentiment polarity of emoji (associated with words) in the same post [17]. It could be argued that it is unlikely that people will mix negative and positive emoji together in a short post. When the emoji strongly reflects the sentiment polarity of words *(emoji indication)*, the indication would serve as a vulnerable point of entry by the social engineer. The following Table 1 presents emotional sentiments weighted against emoji consistency. Table 1 reflects a two way polarity classification (positive against negative sentiment emoji). Emotional inconsistency could result when an emoji is applied wrongly and doesn't reflect sentimental polarity of the words expressed.

Table 1. Emoji indication of consistency *vis-a-vis* inconsistency

lexicon	Word-text	Possible Sentiment	Possible Emoji Consistency	Possible Emoji Inconsistency
Positive	*My boss has left*	*[joy, relief]*		
Negative	*Redo the work*	*[surprised, angry]*		

Emoji intensifiers. SM applications such as *WhatsApp*™ has created new intensifier features that allow the user to project emotional intensity by; (a) the size of the emoji, or (b) the number count of same emoji used.

Table 2. Emoji intensifier

lexicon	Word-text	Level of emotional intensification by size	Level of emotional intensification by number
Positive	*"I'm sooo happy"*		

Social Engineering through Emotional Persuasion. There have been discussions by social engineers about using techniques found in marketing to persuade and gain a victim's trust and compliance [15]. The prize of the social engineer is when the victim is ultimately coerced or persuaded *using in this case emoji* (emotions) to provide sensitive information. In the case of information security, the use of emoji consistency and emoji intensity could be used by social engineers for this purpose. In many cases the social engineers (attackers) will not come into actual contact with potential target victims but will rely on SM applications and most probably use emoji for persuasion. The Elaboration Likelihood Model (ELM) [4] is a possible framework social engineers could use to persuade the victims using emoji to comply (examples are restricted to emoji data) using peripheral route persuasion (indirectly confronting a victim). Cialdini identifies six approaches that social engineers could apply in peripheral route persuasion [18]. As shown by Table 3, these six approaches include; normative commitment; continuance commitment; affective commitment; trust and reactance.

Researchers have begun collecting data which rely on emotion for understanding sentiments [19, 20]. Most have limited the collection to text sentimental analysis using a three way polarity classification. From and information systems security perspective, this work however provides a security mitigating framework that specifically focuses

Table 3. Peripheral route persuasion and research model [4]

Approach	**Construct**	Possible to use emoji	**Friend A**	**Friend B** *reply*
Normative commitment	*Reciprocation as obligation*	Yes	*Hi*	*Hello*
Continuance commitment	*Cognitive investment and perceptual consistency*	Yes		
Affective commitment	*Social "proof" as behavioral modeling and conformance*	Yes		
Trust	*Likeability and credibility*	Yes		
Reactance	*Scarcity and impulsivity*	Yes		

on emoji data. The work looks at a two way polarity classification that social engineers would possible exploit.

4 Methodology

4.1 Procedure

A selected university was used as a case study. Social engineering scenarios were constructed and emoji analysis employed at unique scenarios envisioned for this University. Secondary data sets from information security personnel in the organization were analyzed and emphasis was placed on identifying key words/phrases that were popularly used to persuade victims in social engineering attacks. A study was carried out to determine a range of possible emoji and phrases combination that were persuasive and would be used to social engineer victims. Student-actors were enlisted from the university to perpetrate social engineering attacks using emoji ruses. The student actors were given preliminary briefing on the emoji ruses to apply that involved requesting the following sensitive information; *log-in passwords*, *confidential exam results/marks*; *fellow student bank account details* and *Social Security number/Identity Number*. The student-actors were supported by a senior academic in identifying suitable social engineering ruses that could be used. Screen shots of social engineering using emoji were taken by student-actors and sent to researcher for analysis. Table 4 is an example of possible ruses.

4.2 Prepossessing Emoji Data

Two steps were carried out in order to understand how social engineers would analyze emoji data [21]. The first step involved tokenizing different emoji. Different kinds of emoji data were identified (Fig. 1) and were tokenized and treated as distinct emotions. The second step involved normalizing emoji data where informal emotional emoji intensifiers such as large emoji *vis-a-vis* small emoji (Table 2) were taken to mean various level of emotional intensity.

Table 4. Social engineering ruses

Social Engineering Ruses	Possible word/ emoji scenarios
Request for contact information	*What is your telephone number?* 😊😊
Rush process	*Hey, just received an emergency call, please please give me access so that I help* 😭😭😭
Intimidation	*He knows you were absent yesterday... but I'll not tell* 😉
Name dropping	*I just spoke to Prof* [name withheld] *and he mentioned you by name…* 😬😬😬😬
Requesting forbidden access	*I'll keep your password safe... PROMISE*☺☺!

5 Results and Discussion

A series of 20 social engineering attacks were carried out by student-actors. The results emoji data is shown by Table 5.

Table 5. Corpus statistics

Social Engineering Emoji [*Both attacker and response*]	Frequency [n]	Common Lexicon count
Positive	58	[Affection/kiss **18**] [listening **6**], [smile **7**], [wandering **11**] [Mmm **9**] [wink **5**], [thinking **2**],
Negative	12	[shaking **4**] , [punch **2**], [wait **4**] [angry **2**]

From the 20 executed attacks, 3 three attacks to unsuspecting students were successful. A common characteristic for the successful attacks was that the frequency of emoji used was higher in these attacks. An extract of screenshots data taken from student-actor interactions is shown by Fig. 2 (engaging emotions) and Fig. 3 (exploiting emotions).

Data analysis of various screenshot data reveal two important findings as follows. Firstly, that the effective use of emoji is more likely to persuade victims because of unsuspecting emotional appeal. More time was taken to persuade a victim when only textual words were used in the interaction process.

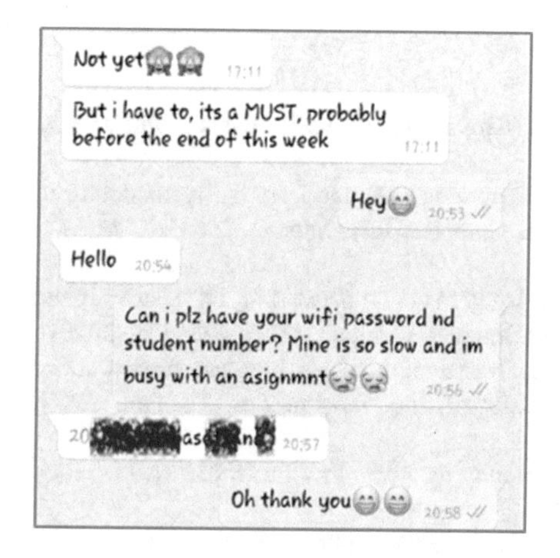

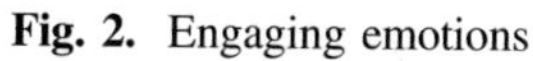
Fig. 2. Engaging emotions

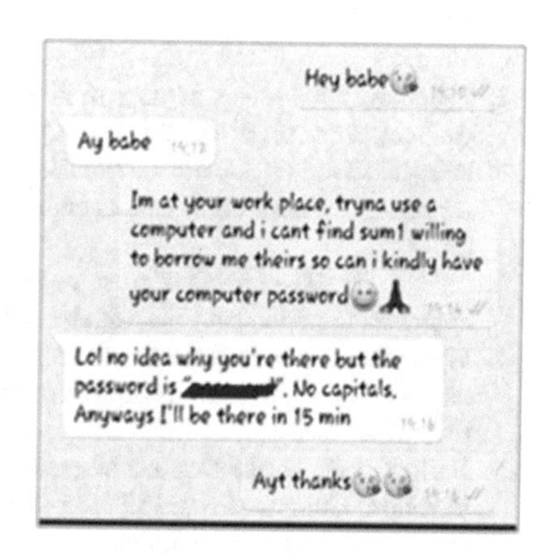

Fig. 3. Exploiting emotions

6 Implications to Information Security Practitioners

Results suggests that emotions and specifically selective use of emoji play a bigger role in persuasion (social engineering) with likely hood of the social engineer escalating attacks using emoji. Successful use of emoji (attacks) was via enforced reciprocity, where the continuous use of emoji 'forced' the unsuspecting victim to 'reciprocate' emotions. The unsuspecting victim tended to provide more information (escalating request) as more was asked. This is in line with similar findings by Workman's research of phishing techniques for social engineering [4]. What contrasts this work from that is the quality and typology used in this work that suggest that a combination of carefully selected emoji will tend to execute escalated attacks much more efficiently.

7 Conclusion

This study was premised in the discipline of information security while also borrowing from disciplines of psychology and criminology. An analysis of the potential for anthropomorphic emoji to be used by social engineers as conduits for escalated attacked to unsuspecting victims was elucidated and carried out. Student-actors social engineered under strictly controlled settings. It was revealed that attacks could be manipulated to desired ends with emotions playing a bigger part in these attacks. The work aims a sensitizing security professionals on the role of emotions and emoji in the domain of security. It is anticipated that this work achieves this purpose.

References

1. Zuber, M.: A survey of data mining techniques for social network analysis. Int. J. Res. Comput. Eng. Electron. **3**(6) (2014)
2. Pohl, H., Domin, C., Rohs, M.: Beyond just text: semantic emoji similarity modeling to support expressive communication. ACM Trans. Comput. Hum. Interact. (TOCHI) **24**(1), 6 (2017)
3. Shin, D.-H.: The effects of trust, security and privacy in social networking: a security-based approach to understand the pattern of adoption. Interact. Comput. **22**(5), 428–438 (2010)
4. Workman, M.: Wisecrackers: a theory-grounded investigation of phishing and pretext social engineering threats to information security. J. Am. Soc. Inform. Sci. Technol. **59**(4), 662–674 (2008)
5. Barlow, J.B., et al.: Don't make excuses! Discouraging neutralization to reduce IT policy violation. Comput. Secur. **39**, 145–159 (2013)
6. Browne, S., Lang, M., Golden, W.: The insider threat-understanding the aberrant thinking of the rogue"trusted agent". In: ECIS (2015)
7. D'Arcy, J., Hovav, A., Galletta, D.: User awareness of security countermeasures and its impact on information systems misuse: a deterrence approach. Inf. Syst. Res. **20**(1), 79–98 (2009)
8. Bulgurcu, B., Cavusoglu, H., Benbasat, I.: Information security policy compliance: an empirical study of rationality-based beliefs and information security awareness. MIS Q. **34** (3), 523–548 (2010)
9. Warkentin, M., Malimage, N., Malimage, K.: Impact of Protection motivation and deterrence on IS security policy compliance: a multi-cultural view. In: Proceedings of the Pre-ICIS Workshop on Information Security and Privacy, Orlando, Paper (2012)
10. DiSalvo, C., Gemperle, F.: From seduction to fulfillment: the use of anthropomorphic form in design. In: Proceedings of the 2003 International Conference on Designing Pleasurable Products and Interfaces. ACM (2003)
11. Mogicons. Emoticons for Facebook (2017). https://www.mogicons.com/en/. Accessed 5 May 2017
12. Mitnick, K.D., Simon, W.L.: The Art of Deception: Controlling the Human Element of Security. Wiley, New York (2011)
13. Granger, S.: Social engineering fundamentals, part I: hacker tactics. SecurityFocus **18** (2001)
14. Palumbo, J.: Social engineering: what is it, why is so little said about it and what can be done? SANS Institute (2000)
15. Gao, W., Kim, J.: Robbing the cradle is like taking candy from a baby. In: Proceedings of the Annual Conference of the Security Policy Institute (GCSPI) (2007)
16. Calluzzo, V.J., Cante, C.J.: Ethics in information technology and software use. J. Bus. Ethics **51**(3), 301–312 (2004)
17. Hu, X., et al.: Unsupervised sentiment analysis with emotional signals. In: Proceedings of the 22nd international conference on World Wide Web. ACM (2013)
18. Cialdini, R.B.: Science and Practice (2001)
19. Pak, A., Paroubek, P.: Twitter as a corpus for sentiment analysis and opinion mining. In: LREC (2010)
20. Bifet, A., Frank, E.: Sentiment knowledge discovery in twitter streaming data. In: International Conference on Discovery Science. Springer, Heidelberg (2010)
21. Kouloumpis, E., Wilson, T., Moore, J.D.: Twitter sentiment analysis: the good the bad and the omg! ICWSM **11**(538–541), 164 (2011)

Personal Data Protection Act Enforcement with PETs Adoption: An Exploratory Study on Employees' Working Process Change

May Fen Gan, Hui Na Chua(✉), and Siew Fan Wong

Department of Computing and Information Systems, Sunway University, Subang Jaya, Malaysia
huinac@sunway.edu.my, siewfan@gmail.com

Abstract. It is often that personal data were being misused by organizations for their own benefits. To tackle this issue, different countries had introduced and enforced personal data protection regulations. With the enforcement, organizations in the relevant countries need to comply with the law enforcement to protect personal data as their legal responsibility. Privacy Enhancing Technologies (PETs) act as a form of technology that protects individual privacy data in organizations. The purpose of this research is to discover the impact of personal data protection act enforcement with PETs adoption on organization employees' working experience and performance through the study of their working process change. This research adopts a qualitative single case study on one of the telecommunication companies in Malaysia. The targeted participants are employees come from different work nature, i.e., use personal data, process personal data or setup system to protect personal data. The finding of this research will enable organizations to have better understanding in future PETs adoption and provide insights on the measures to be taken to comply with personal data protection. This paper presents our preliminary results based on semi-structured interviews with 8 participants from different groups of work nature.

Keywords: Privacy enhancing technologies · Technology adoption · Personal data protection · Information privacy

1 Introduction

Users provide their personal information to different company service providers including banking, healthcare and telecommunications in exchange for their services. To ensure users' personal data not to be misused, it is important to have the personal information well protected by these companies. Each company has their respective information privacy policy to protect users' personal information. On top of that, the information privacy policy must comply with the data protection principles. Different countries had enacted relevant acts in protecting personal information. For instance, UK Data Protection Act 1998 (DPA), European General Data Protection Regulation (GDPR) and The Federal Trade Commission Act (FTCA). Without exception, Malaysia government had enforced Personal Data Protection Act (PDPA) in 2013 to protect individual's personal data in any commercial transactions [1].

K.J. Kim et al. (eds.), *IT Convergence and Security 2017*,
Lecture Notes in Electrical Engineering 450,
DOI 10.1007/978-981-10-6454-8_25

Many companies have adopted Privacy Enhancing Technologies (PETs) to comply with their respective country data protection act with ease. PETs refer to any technologies that are used to protect personal data. PETs have been the center of attention in many countries, especially the USA, where the legislative provisions for data protection are efficient to protect personal data. It is better thought of as complementary to law with which they must work together to provide a robust form of privacy protection [2].

In Malaysia, recent study indicates the adoption of PDPA is affecting organization working processes among employees [3]. A number of studies have been conducted in term of technology adoption affecting working processes. Positive and negative impacts have been shown in these research [3–5]. However, prior literature shows limited understanding of how the PETs adoption impacts employees' working processes which inevitably influencing their experience (such as acceptance and perception) and work performance (such as efficiency and effectiveness) [6, 7]. Hence, this research aims to fill in this gap. Telecommunication industry is selected as the study of this research as it is a sector that collects handles and process large number of data in a daily basis. Thus, this research seeks to answer the following research questions:

1. What kind of working processes is affected in adopting PETs?
2. How do employees experience the change and adapt with the new system environment?
3. What is the impact on employees' performance in adopting PETs?

By analyzing and answering these questions, this research will provide insights on the impact of employees' performance and experience on PDPA enforcement with PETs adoption. This allows organization to have better strategic planning in future PETs adoption, employees' working processes and procedures for personal data protection. Besides that, this research provides an insight on the measures taken to comply with PDPA.

This paper presents a preliminary study results of exploring employees' working processes in adopting PETs after PDPA enforcement. A single case qualitative approach has been taken in this research. This preliminary result is based on the semi-structured interview of a total of 8 participants from different departments.

2 Literature Review

2.1 Protection of Information Privacy

Information privacy has been studied extensively over the years, but there is no one standard definition for information privacy. Different researchers have defined information privacy in different ways. However, they do share a common goal which is individual has the right to control and access their own data. Table 1 shows different definition by various researchers.

There might be misuse of personal information for ones' own benefit. For example, selling customer data to a third party. This leads to privacy invasions. Privacy invasions can be divided into misappropriation, intrusion upon seclusion, false light and public

Table 1. Privacy definitions

Author(s)	Definition
Westin (1967)	The claim of individuals, groups and institutions to determine for themselves, when, how and to what extent information about them is communicated to others
Stone et al. (1983)	The ability (i.e., capacity) of the individual to control personally (vis-à-vis other individuals, groups, organizations, etc.) information about one's self
Smith (1993)	A condition of limited access to identifiable information about individuals
Clarke (1999)	The interest an individual has in controlling, or at least significantly influencing, the handling of data about themselves
Bélanger and Crossler (2011)	The desire of individuals to control or have some influence over data about themselves

disclosure of private facts. For example, one victim suffers from monetary loss when he clicks on an internet link which is a malware virus [8].

This leads to individual' concern of their personal data and privacy information being misused. Individuals will be concerned for their privacy if there is insufficient trust between user and organization [9]. However, individuals have little knowledge or lack of control on their own information that maintained by organization's database and other sector entities [10].

The government involvement to protect individuals' privacy by establishing legislative and regulatory methods may be one of the method in handling privacy concerns. Besides that, personal data can be protected through self-regulations, privacy education and the utilization of Privacy Enhancing Technologies (PETs) [11].

2.2 Privacy Enhancing Technologies

The European Commission defines PETs as "a coherent system of ICT measures that protects privacy by eliminating or reducing personal data or by preventing unnecessary and/or desired processing of personal data, all without losing the functionality of the information system" [12]. It is usually referring to the use of technology to help achieve compliance with data protection legislation [13].

Different PETs were developed for various data protection purposes which may include either both privacy and security protection features. For any protection that involves technology such as collection or processing of the data, regardless of personal or non-personal, it will belong to PETs [14]. Thus, security and privacy protection on technology falls under PETs.

Olivier [15] structures PETs in four different layers which are personal privacy enhancing technologies, web-based technologies, information brokers, network-based technologies. In each layer, different technologies are used to protect data. For in-stance, cookie managers, ad-blockers, encryption software, privacy networks, firewalls, etc. PETs are not limited to certain specific techniques only and it might not necessary to be something new or something that has never see or done before [2].

Even though PETs are useful in protecting data, there are several obstacles in adopting it. Borking [16] shows three obstacles in adopting PETs. Firstly, lack of availability of PETs and lack of user friendliness. Secondly, it is insufficiently supported by current regulations. Thirdly, there are major obstacles towards deployment infrastructures. To overcome these obstacle, it is important to understand the structure and properties of PETs. Goldberg [17] suggested a list of technologies properties that are required for PETs to be useful. For instance, usability, deployability, effectiveness and robustness.

2.3 Impact of Technology Adoption on Work Processes

Organizations adopt technology to enhance their competitiveness as well as to increase the efficiency and effectiveness of processes and products while lowering the operating cost [6]. For instance, finding shows that there is a positive relationship between technology adoption and hospital communication performance [5]. There are certain factors that will affect the technology adoption such as relative advantage, compatibility, complexity, top management support, firm size, technological competence and competitive pressure [18]. In addition, obstacles in adopting new technology includes the resistance to change, uncertainty of new technology and employees' acceptance of change [7].

The introduction of a communication tool – eWhiteboard was introduced for the support of inter-team coordination of patient status within the surgical flow. It is a platform for care providers to communicate and share the care status and distribute information on specific patient needs. After implementing this tool for 8 months, observation shows that the communication load among care providers has reduced. Thus, the interruptions to clinical work is reduced as there is lower incidence of phone calls or face to face interruption [5].

In a study of personal data protection policy adoption, there is an indication shows that adopting personal data protection slows down the organization process as there are more procedure needs to be followed [3].

3 Problem Statement

In Malaysia, PDPA is an enforcement on organizations to protects users' personal data in any commercial transactions [1]. Therefore, it is a legal responsibility of organizations to comply with the act to ensure personal data of their customers is protected. With the adoption of PETs, personal data is expected to be better protected. However, there is a case where millions of personal information records have been stolen or lost in the year 2015 [19].

Researchers have identified the factors and importance in adopting new technologies [18, 20]. However, prior literature shows limited study on how PETs adoption affect employees' working processes after PDPA enforcement and how PETs adoption influence employees' experience and performance. By analyzing these aspects, organization is able to have better understanding and subsequently planning for more strategic PETs adoption. Further, the management team can utilize the finding as guidelines to improve the organization's procedures for personal data protection. In addition, organization can optimize the process of PETs adoption in the future by understanding employees' experience and how this influence their work performance in the current processes.

4 Methodology

Qualitative single case study is adopted in this research to investigate the research questions. There are a number of reasons in choosing qualitative case study research. First, qualitative research provides both depth and detail in the responses, rather than attempting to fit the experiences of individuals into pre-determined answers. Moreover, the use of case study is to answer "How" and "Why" questions [21]. Thus, it is suitable to be adopted in this research since the research questions has "How" and "Why" elements.

In addition, an exploratory study is adopted when there is limited information available and researchers have little control over the event [21]. Moreover, this approach is chosen as the investigations area and topic are less understood and have been less investigated in the past [22]. In addition, researchers are able to understand the nature and complexity of the process that is taking place and gain an in-depth understanding of the phenomenon under study through case study [4].

4.1 Data Collection

In this research, participants' interview data is collected from one of the telecommunication industries in Malaysia. The participants were selected based on purposeful sampling. It is based on the purpose of the study which is to discover the impact on employees' experience and performance through the understanding of their work process change. We identified three groups of users that have different roles in the whole process of data request, process and dissemination. These groups of users are:

i. Data User – A user who requests data from data processor to perform their work such as customer profile understanding and marketing.
ii. Data processor – A user who has the access to the database and personal information. Data processor processes personal data based on the requests from data user.
iii. Data Controller – A user who is responsible in defining data protection procedures and approving data requests. Any request from data user will go through data controller for verifying the purpose of the request.

We conducted semi-structured interviews with open-ended questions, which allow us to follow up any unexpected responses. During the interviews, participants are free to elaborate on the responses based on their perception. A total of 8 participants have been participated in this interview thus far. Each interview took between 65 min to 90 min. Data confidentiality and anonymity of respondents were guaranteed. All interviews will be audio recorded for analysis purposes.

4.2 Reliability and Validity

Trustworthiness of qualitative research can be established using the terms credibility (validity), dependability (reliability), confirmability and transferability [23]. To ensure data reliability and validity, several procedures have been followed. Firstly, all recorded interviews will be transcribed for analysis. Secondly, a rich description in conveying the findings is provided. Thirdly, discrepant information identified is presented in the report as well [24]. Lastly, all the research will be documented [21].

5 Research Findings and Analysis

This research refers to a conceptual model of implementation research by Proctor [25]. The model shows the strategies in implementation and the outcomes. Besides that, there are several aspects in affecting implementation [26]. In addition, [27] shows several components for policy implementation research. The preliminary result of this exploratory research is presented in the following subsections, and our analysis rationale is guided partially by the aspects discussed in [25–27].

5.1 Comparing Work Process Flow

Different groups of users have different work processes. For data users, as presented in Fig. 1 shows that there are three significant changes before and after PDPA enforcement with PETs adoption. First, it shows that data users need to go through more process layers in getting approval. After PETs adoption, there is an introduction of security team (i.e., data controller). The purpose of security team is to validate and filter the data requests. Hence, the process of requesting data is more tightened as before. In addition, as the process flow of requesting data increases, data users might delay in executing their tasks. Second, the company had introduced an additional computerized system in data request procedure, a.k.a, IT CR system. Third, the access control of the system. Each user will have their own username and password to log in to the system. This allows data controller to track data users' activities in the system.

For data processor users, as presented in Fig. 2, an important takeaway in data processor work flow is the methods used to send data to data users. Before PETs adoption, they can send it via email, shared folder or external hard disk. According to the interviewee, the external hard disk can be either a personal hard disk or company hard disk. This increases the chance of privacy risk. After year 2013, the company improved the method of sending data is by only allowing data processor to send the data through shared folder and secure email. However, it is untraceable on how the user uses the obtained data. Besides that, data processor received less user request due to the filtering process by security team. Previously, data users can easily obtain any data from data processors if the data request is approved only by their supervisor. After the PDPA enforcement with PETs adoption, there is a constraint which security team will evaluate the needs of the data request.

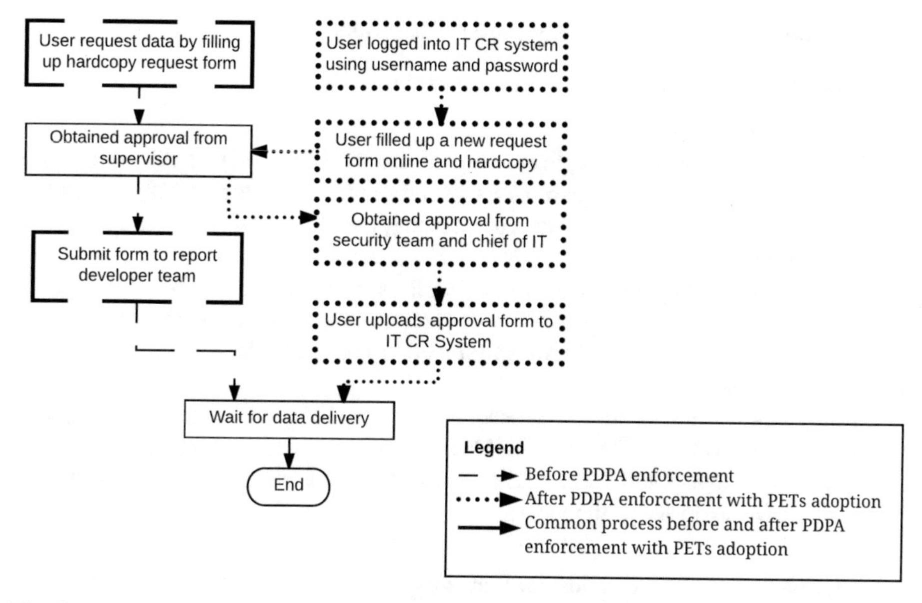

Fig. 1. A summarized data user workflow – before and after PDPA enforcement with PETs adoption

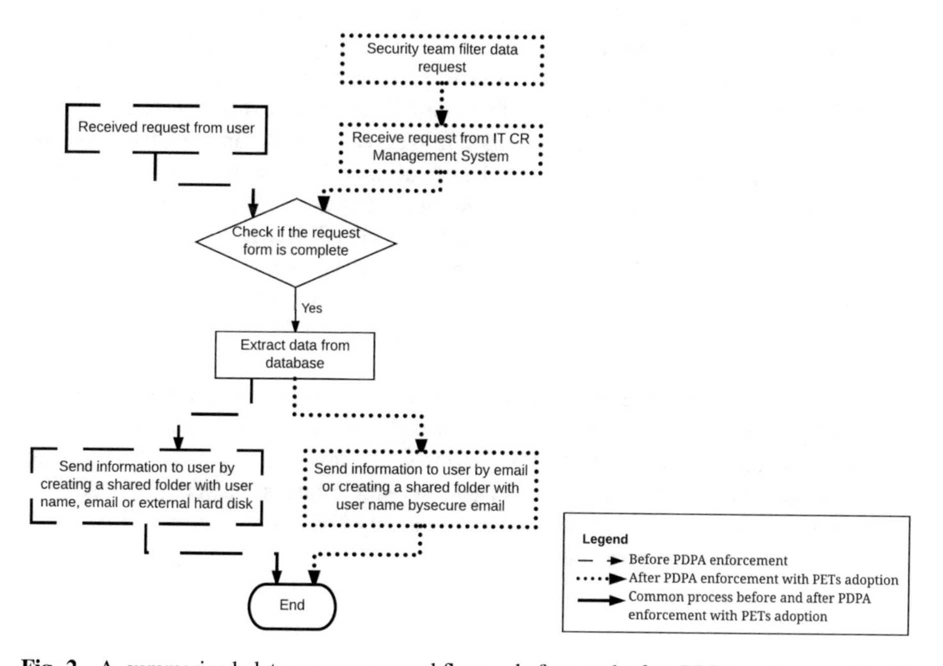

Fig. 2. A summarized data processor workflow – before and after PDPA enforcement with PETs adoption

5.2 Employees' Change and Adaptation Experience

The following present themes identified based on the interviews.

- The Awareness of Change
  Only small numbers of employees are aware of PDPA via email notification. Addition to that, they experienced PET adoption only when their superior informed them. Usually, meeting is conducted regarding the change of the system.
- Perception of Data Protection
  Although the participants think that PETs can protect personal information, but it is not able to fully protect personal information. Interviewees think that personal compliance is an important factor in protecting personal information.
- Technology used
  There are numerous technologies used in adopting PETs which include encryption, Multicast File Transfer Protocol, SSH File Transfer Protocol and introducing of new systems. Employees think that PETs are useful in protecting data and there are no major problems in using the system.

5.3 Impact of PETs Adoption on Performance

The following shows the impacts of PETs adoption after PDPA enforcement on participants' work performance.

- Efficiency
  The workload for data users increases after PETs adoption because there are more procedures that need to be followed. For example, they need to get additional approval from security team and chief of IT. This delays the process of getting the data and subsequently delaying the user to perform any analysis on the data.
  On the other hand, data processors think that their workload has been decreased. Before PETs adoption, data processors will receive a lot of data requests from the data users. With the deployment of data controllers' (security team) procedures, which filter the requests from data users which directly decreasing the number of requests. This filtering process lowers the burden of data processors. Therefore, with the deployment of security team's procedures for personal data protection; it decreases data processors workload and essentially tightening the process of data request.
- Training given
  As mentioned earlier, the new system is adopted due to the enforcement of PDPA. Training is provided when *new system is introduced*. However, there is no in-depth introduction in understanding PDPA. During the yearly security meeting, only brief introduction on PDPA is conducted. Therefore, numerous employees still unsure of what is PDPA. For instance, some employees only know the purpose of PDPA but they do not know the principles of PDPA.

6 Conclusion

This preliminary result shows that different groups of participants experience different impacts in adopting PETs after PDPA enforcement. The impacts are varied based on the nature of work. It is also observed that all participants have positive perception on PETs adoption for personal data protection, and additionally, personal compliance to the data protection policy is perceived a critical factor for successful implementation.

Our preliminary result thus far presents the semi-structured interview outcomes from data user and data processor perspectives. We will continue this research to gain more complete understanding from additional data controller perspective. We aim to publish the future work results according to all the different groups (i.e., data user, data processor and controller) perspectives and use various methods to perform qualitative data analysis.

Acknowledgement. This research is supported by the Malaysian Government FRGS Grant (FRGS/1/2015/SS03/SYUC/02/1).

References

1. Personal Data Protection Act 2010.: Malaysia Attorney General's Chambers (2010)
2. Blarkom, G.W.V., Borking, J.J., Verhaar, P.: PET. Handbook of privacy and privacy-enhancing technologies. In: Blarkom, G.W.V., Borking, J.J., Olk, J.G.E. (eds.) Handbook of Privacy and Privacy-Enhancing Technologies: The case of Intelligent Software Agents, Cambridge (2003)
3. Chua, H.N., Wong, S.F., Chang, Y.H., Tan, C.M., Gan, M.F.: Personal data protection policy adoption: an exploratory study of Telco Industry's Implementation. In: 21st Biennial Conference of the International Telecommunications Society, Taipei, 26–29 June 2016
4. Cao, Q., Jones, D.R., Sheng, H.: Contained nomadic information environments: technology, organization, and environment influences on adoption of hospital RFID patient tracking. Inf. Manage. **51**(2), 225–239 (2014)
5. Taneva, S., Law, E., Higgins, J., Easty, A., Plattner, B.: Operating room coordination with the eWhiteboard: the fine line between successful and challenged technology adoption. Health Technol. **1**(2–4), 81–92 (2011)
6. Aboelmaged, M.G.: Predicting e-readiness at firm-level: an analysis of technological, organizational and environmental (TOE) effects on e-maintenance readiness in manufacturing firms. Int. J. Inf. Manage. **34**(5), 639–651 (2014)
7. Delaney, R., D'Agostino, R.: The Challenges of Integrating New Technology into an Organization. Mathematics and Computer Science Capstones, 25 (2015)
8. Zolkepli, F.: 'Zeus' Targets Mobile Phone User. The Star Online (2014). http://www.thestar.com.my/News/Nation/2014/09/25/Zeus-targets-mobile-phone-users-Scammers-rake-in-thousands-through-malware-virus
9. Bergström, A.: Online privacy concerns: a broad approach to understanding the concerns of different groups for different uses. Comput. Hum. Behav. **53**, 419–426 (2015)
10. Gellman, R.: The digital person: technology and privacy in the information age. Gov. Inf. Q. **22**(3), 530–532 (2005)
11. Seničar, V., Jerman-Blažič, B., Klobučar, T.: Privacy-enhancing technologies—approaches and development. Comput. Stand. Interfaces **25**(2), 147–158 (2003)

12. London Economics.: Study on the Economic Benefits of Privacy-Enhancing Technologies (PETs) (2010). http://ec.europa.eu/justice/policies/privacy/docs/studies/final_report_pets_16_07_10_en.pdf
13. Kenny, S.: An Introduction to Privacy Enhancing Technologies (2008). https://iapp.org/news/a/2008-05-introduction-to-privacy-enhancing-technologies/
14. Chan, J.M.J., Chua, H.N., Lee, H.S., Iranmanesh, V.: Privacy and security: how to differentiate them using privacy-security tree (PST) classification. In: International Conference on Information Science and Security (2016)
15. Olivier, M.S.: A layered architecture for privacy-enhancing technologies. S. Afr. Comput. J. **31**, 53–61 (2003)
16. Borking, J.J.: Why adopting privacy enhancing technologies (PETs) takes so much time. In: Gutwirth, S., Poullet, Y., De Hert, P., Leenes, R. (eds.) Computers, Privacy and Data Protection: An Element of Choice, pp. 309–341. Springer, Dordrecht (2011)
17. Goldberg, I.: Privacy Enhancing Technologies for the Internet III: Ten Years Later (2007). http://www.cypherpunks.ca/~iang/pubs/pet3.pdf
18. Wang, Y.-S., Li, H.-T., Li, C.-R., Zhang, D.-Z.: Factors affecting hotels' adoption of mobile reservation systems: a technology-organization-environment framework. Tour. Manag. **53**, 163–172 (2016)
19. MalaysianDigest: Cybersecurity Threats: The Risk is Real in Malaysia (2016). http://www.malaysiandigest.com/technology/605954-cybersecurity-threats-the-risk-is-real-in-malaysia.html
20. Weerd, I.V.D., Mangula, I.S., Brinkkemper, S.: Adoption of software as a service in Indonesia: examining the influence of organizational factors. Inf. Manag. **53**(7), 915–928 (2016)
21. Yin, R.K.: Case Study Research: Design and Methods, 5th edn. SAGE Publications, Thousand Oaks (2009)
22. McGivern, Y.: The Practice of Market and Social Research: An Introduction. Pearson Education, Essex (2007)
23. Bloomberg, L.D., Volpe, M.: Completing Your Qualitative Dissertation. Sage Publications, Thousand Oaks (2012)
24. Creswell, J.W.: Research Design: Qualitative, Quantitative, and Mixed Methods Approaches. Sage Publications, Thousand Oaks (2013)
25. Proctor, E.K., Landsverk, J., Aarons, G., Chambers, D., Glisson, C., Mittman, B.: Implementation research in mental health services: an emerging science with conceptual, methodological, and training challenges. Adm. Policy Ment. Health Ment. Health Serv. Res. **36**(1), 24–34 (2009)
26. Durlak, J.A., DuPre, E.P.: Implementation matters: a review of research on the influence of implementation on program outcomes and the factors affecting implementation. Am. J. Community Psychol. **41**(3–4), 327 (2008)
27. Moulton, S., Sandfort, J.R.: The strategic action field framework for policy implementation research. Policy Stud. J. **45**(1), 144–169 (2017)

An Improved Iris Segmentation Technique Using Circular Hough Transform

Kennedy Okokpujie(✉), Etinosa Noma-Osaghae, Samuel John, and Akachukwu Ajulibe

Department of Electrical and Information Engineering, College of Engineering, Covenant University, Ota, Ogun State, Nigeria
{kennedy.okokpujie,etinosa.noma-osaghae, samuel.john}@covenantuniversity.edu.ng, akachi_benji@yahoo.com

Abstract. It is quite easy to spoof an automated iris recognition system using fake iris such as paper print and artificial lens. False Rejection Rate (FRR) and False Acceptance Rate (FAR) of a specific approach can be as a result of noise introduced in the segmentation process. Special attention has not been paid to a modified system in which a more accurate segmentation process is applied to an already existing efficient algorithm thereby increasing the overall reliability and accuracy of iris recognition. In this work an improvement of the already existing wavelet packet decomposition for iris recognition with a Correct Classification Rate (CCR) of 98.375% is proposed. It involves changing the segmentation technique used for this implementation from the integro-differential operator approach (John Daugman's model) to the Hough transform (Wilde's model). This research extensively compared the two segmentation techniques to show which is better in the implementation of the wavelet packet decomposition. Implementation of the integro-differential approach to segmentation showed an accuracy of 91.39% while the Hough Transform approach showed an accuracy of 93.06%. This result indicates that the integration of the Hough Transform into any open source iris recognition module can offer as much as a 1.67% improved accuracy due to improvement in its preprocessing stage. The improved iris segmentation technique using Hough Transform has an overall CCR of 100%.

Keywords: Integro-differential operator · Segmentation · Wave packet decomposition · False Rejection Rate (FRR) · Hough transform · False Acceptance Rate (FAR) · Recognition Accuracy (RA)

1 Introduction

Increased demand for more trustworthy security systems has led to the application of biometric security systems in various ways [1, 5]. When individuals are automatically recognized, based on their physiological or behavioural characteristics, biometrics is the base parameter in use. The fingerprints, voice and iris are some major examples of biometrics and they have a wide range of application areas [6, 7]. The Iris is a very accurate biometric parameter that is not susceptible to the aging effect.

K.J. Kim et al. (eds.), *IT Convergence and Security 2017*,
Lecture Notes in Electrical Engineering 450,
DOI 10.1007/978-981-10-6454-8_26

The Iris Biometric Recognition System can be spoofed with fakes such as artificial iris etc. An Iris biometric recognition system that cannot be spoofed easily drastically increases the trust placed on it by its users.

Daugman introduced the use Fast Fourier Transform (FFT) to check the iris pattern [2]. In his proposition, the spectrum in high frequency domain was used to differentiate one iris pattern from the other. Although a purposely blurred and defocused fake iris may be falsely accepted by the iris recognition system.

In this work, an improvement of the already existing wavelet packet decomposition for iris recognition is provided. It involves changing the segmentation technique used for this implementation from the integro-differential operator approach (John Daugman's model) to the Hough transform approach (Wilde's model).

The objectives of the study include, implementing a new iris segmentation algorithm to build a more robust iris recognition algorithm, designing a flowchart for the implementation of the proposed algorithm and using MATLAB to analyze the results.

The overall aim of this study is basically the implementation of a new segmentation technique on Hough Transform to build an improved wavelet decomposition algorithm for authentication using iris recognition.

2 Background and Literature Review

2.1 The Circular Hough Transform

The Circular Hough Transform is used to locate any regular curve in a given geometric shape, or shapes in a given image. It redefines the image as forms of ellipses, circles and expressions with powers of three and above. Circular Hough Transform was used to localize irises by Wildes et al. [3]. Wildes proposed the generation of the points of the parametric form by computing the initial derivatives of the image's brightness and thresholding the resulting values. Hough transform techniques have some drawbacks. First, threshold values are required for tracing out the parametric form, and doing away with important points in the image can lead to the formation of a poor image template. The Hough transform needs a large memory and special hardware for its computation. This makes it expensive for real time applications.

2.2 The Integro-Differential Operator

The Integro-differential operator was the brainchild of Daugman who used it to detect the parametric properties of the iris [4]. It makes direct use of the differential derivations and does not do well in removing noise from the image template formed. But it is not encumbered with the thresholding problem of the Circular Hough Transform.

3 Methodology and Proposed Framework Data Flow

MATLAB® was used to evaluate the Daugman (integro-differential) and Wildes (Circular Hough) methodologies respectively.

The framework data flow consists of the following as shown in Fig. 1 Database, load image, segmentation algorithm, normalization algorithm, feature extraction algorithm and matching algorithm.

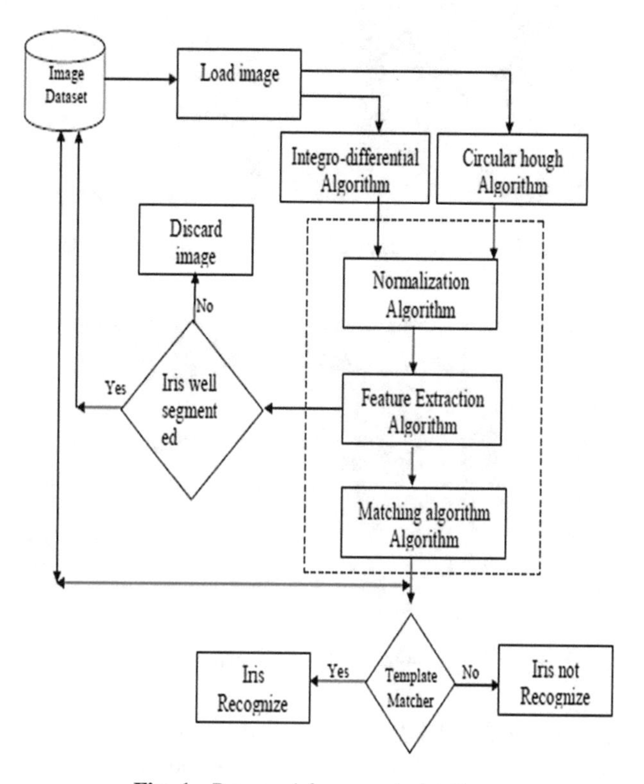

Fig. 1. Proposed framework dataflow

4 Database Collection

To test the developed system, some set of iris data from the Chinese Academy of Sciences - Institute of Automation (CASIA) eye image database were used. CASIA Iris Image Database includes 1080 iris images from 108 eyes. For each eye, 10 images are captured in two sessions with a self-developed CASIA close-up iris camera, where five samples are collected in the first session and five in the second session. All images are stored as BMP format with 320×280 resolution. The CASIA image dataset used contains 6 subjects and 10 different images of each unique eye.

5 Implementation and Validation

Each eye image tested was selected and run through the simulated program and all the processes involved in iris segmentation were carried out. The intra-class and inter-class matching was carried out and recorded in a tabular form, from where further analyses

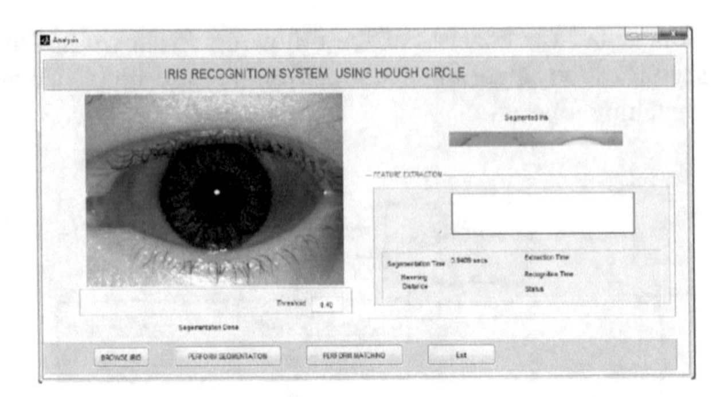

Fig. 2. GUI for circular Hough segmentation process

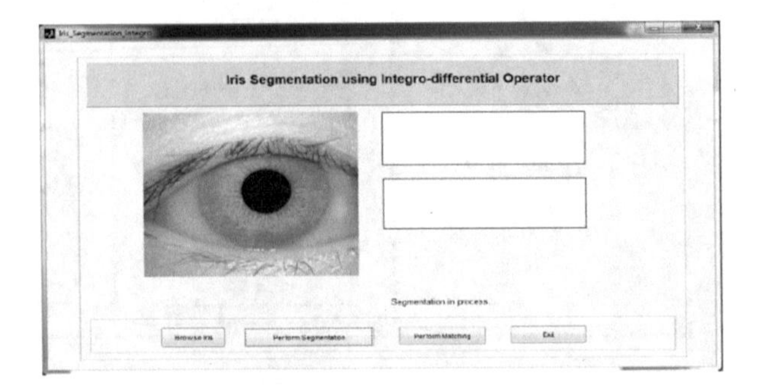

Fig. 3. GUI for integro-differential segmentation process

were carried out. Figures 2 and 3 display screen shots of the application interfaces that performed the recognition process for each algorithm on the selected eye image.

After the image was loaded, segmentation of the selected eye image using circular Hough transform and integro-differential operator algorithms were carried out. After segmentation, the program also performed normalization and feature extraction on the iris image using Daugman's rubber sheet and Log Gabor algorithms respectively.

6 Result and Discussion

The result of the intra-class matching using the integro-differential operator is displayed on (Table 1). Eye image class one and two had an error of 8.33%, eye image class three and four had an error of 10%, eye image class five had an error of 5.83% and eye image class six an error of 9.17%. The result of the intra-class matching using the circular Hough transform is displayed on (Table 2). Eye image class one and two both recorded an error of 7.5%, eye image class three had an error of 5.83%, eye image class four had an error of 8.33%, eye image class five had an error of 5% and eye image class six had an error of 7.5%. Running the intra-class for each method shows zero percent False Acceptance Error (FAR) rate.

Table 1. FAR and FRR (integro-differential operator)

S/N	FAR	FAR (%)	FRR	FRR (%)	Accuracy (%)
01	0	0	10	8.3333	91.6667
02	0	0	10	8.3333	91.6667
03	0	0	12	10.0000	90.0000
04	0	0	12	10.0000	90.0000
05	0	0	7	5.8333	94.1667
06	0	0	11	9.1667	90.8333

Table 2. FAR and FRR (circular Hough transform)

S/N	FAR	FAR (%)	FRR	FRR (%)	Accuracy (%)
01	0	0	9	7.5000	92.5000
02	0	0	9	7.5000	92.5000
03	0	0	7	5.8333	94.1667
04	0	0	10	8.3333	91.6670
05	0	0	6	5.0000	95.0000
06	0	0	9	7.5000	92.5000

The intra-class and inter-class hamming distance for the integro-differential operator and circular Hough transform implementations are shown from Figs. 4, 5, 6 and 7.

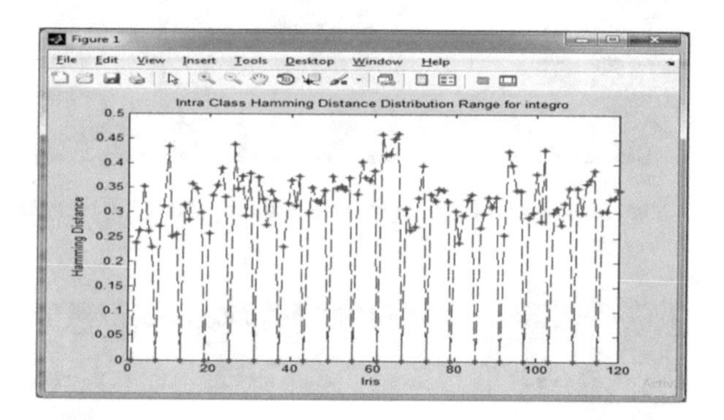

Fig. 4. Intra-class graph for integro-differential operator implementation.

Figures 4 and 5 show the intra-class graph for integro-differential and circular Hough transform respectively. The graphs show the eye images that are rejected because they fall above the threshold value of 0.4. The systems are reliable since their intra-class rejection rates are acceptable.

Figures 6 and 7 show the inter-class graph for integro-differential and circular Hough implementations. From the graphs, eye images were rejected because they fall above the threshold value of 0.4. This shows that the threshold set for the application was acceptable. It also shows that the system is reliable since False Acceptance Error is extremely low.

From each run of the intra-class and inter-class matching, numbers of FRR and FAR and the percentage of error for each run were taken. In addition, the percentage

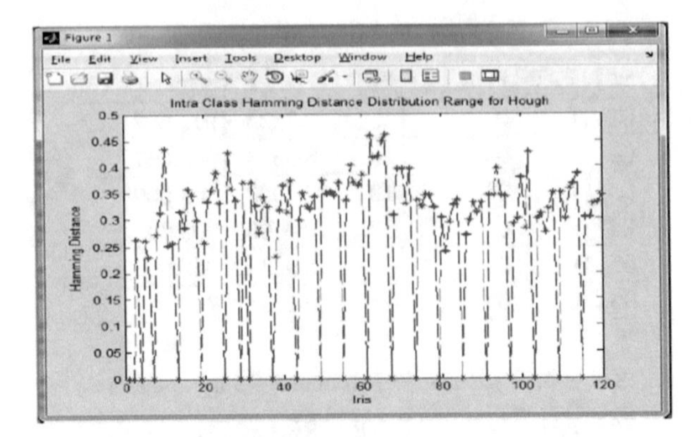

Fig. 5. Intra-class graph for circular Hough transform implementation.

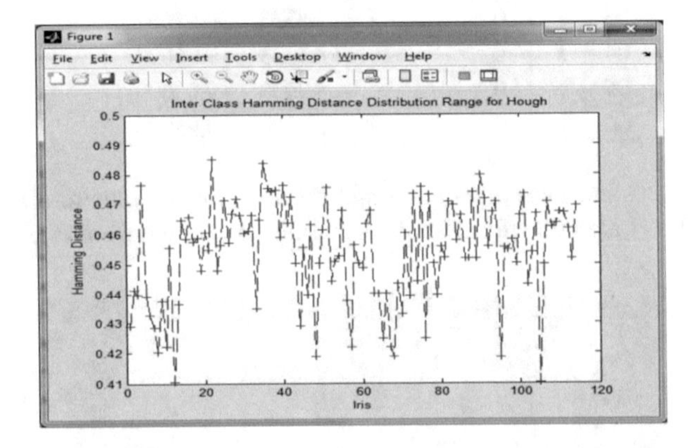

Fig. 6. Inter-class graph for integro-differential operator implementation.

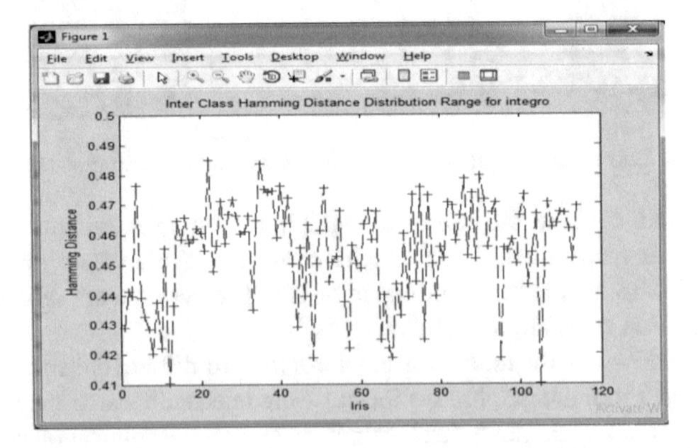

Fig. 7. Inter-class graph for circular Hough transform implementation.

accuracy for each method was calculated. To calculate the value for the percentage recognition accuracy, the value of the FRR and the FAR were subtracted from 100. This is called the Recognition Accuracy (RA). These results are displayed in a tabular form (Tables 1 and 2). The FRR of the integro-differential operator and the circular Hough transform implementations are plotted on a graph and shown below:

Figures 6 and 7 show the inter-class graph for integro-differential and circular Hough implementations. From the graphs, eye images were rejected because they fall above the threshold value of 0.4. This shows that the threshold set for the application was acceptable. It also shows that the system is reliable since False Acceptance Error is extremely low (Fig. 8).

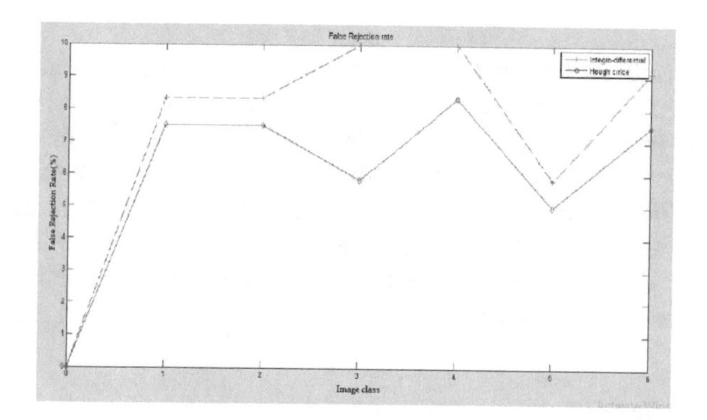

Fig. 8. FRR for Circular Hough and Integro-differential implementations.

From each run of the intra-class and inter-class matching, numbers of FRR and FAR and the percentage of error for each run were taken. In addition, the percentage accuracy for each method was calculated. To calculate the value for the percentage recognition accuracy, the value of the FRR and the FAR were subtracted from 100. This is called the Recognition Accuracy (RA). These results are displayed in a tabular form (Tables 1 and 2). The FRR of the integro-differential operator and the circular Hough transform implementations are plotted on a graph and shown below:

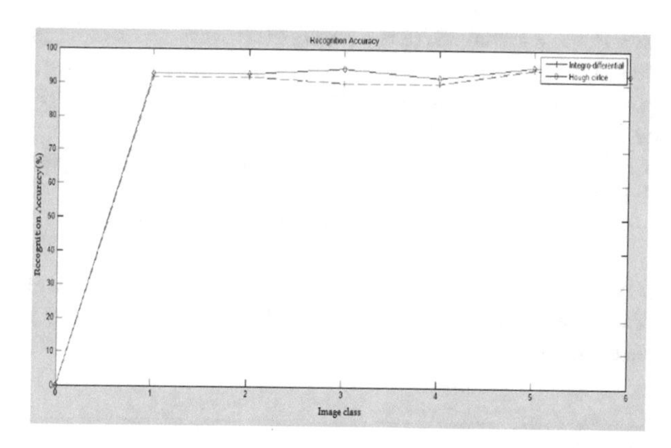

Fig. 9. Recognition accuracy for circular Hough transform and Integro-differential operator implementations

From the graph in Fig. 9, it was observed that the Circular Hough Transform implementation curve was higher than that of the Integro-differential operator implementation. This implies that the circular Hough transform implementation's degree of accuracy is higher than that of the Integro-differential operator implementation's accuracy. This shows that the circular Hough transform implementation is more accurate than the integro-differential transform implementation.

The accuracy of the system is determined by the FAR and FRR. For both implementations, FAR is zero (0).

7 Conclusion

This paper has proved that under the same conditions, Wildes' algorithm performs better than Daugman's algorithm because the FRR curve of circular Hough transform implementation was lower than that of the Integro-differential operator implementation. Also, the Recognition Accuracy (RA) graph of the two algorithms depicts higher accuracy for Wildes over Daugman. Hence, there is basis that if the circular Hough transform proposed by Wildes is used as a segmentation algorithm in the proposed wavelet packet decomposition, it will perform better than the integro-deferential algorithm proposed by Daugman.

8 Future Work

Other forms of segmentation besides Daugman and Wilde would be considered and a very efficient hybrid segmentation technique would be developed.

Acknowledgement. This paper was sponsored by Covenant University, Ota, Ogun State, Nigeria.

References

1. Jain, A.K., Bolle, R.M., Pankanti, S. (eds.): Biometrics: Personal Identification in Networked. Kluwer, Norwell (1999)
2. Daugman, J.: Demodulation by complex-valued wavelets for stochastic pattern recognition. Int. J. Wavelets Multiresolut. Inf. Process. **1**(1), 1–17 (2013)
3. Zang, H., Sun, Z., Tan, T.: Contact Lens Detection Based on Weighted LBP. Chinese Academy of Sciences, Beijing (2010)
4. Wildes, R.P.: Iris recognition: an emerging biometric technology. Proc. IEEE **85**(9), 1348–1363 (1997)
5. Badejo, J.A., Atayero, A.A., Ibiyemi, T.S.: A robust preprocessing algorithm for iris segmentation from low contrast eye images. In: Future Technologies Conference (FTC), pp. 567–576. IEEE (2016)

6. Okokpujie, K., Olajide, F., John, S., Kennedy, C.G.: Implementation of the enhanced fingerprint authentication in the ATM system using ATmega128. In: Proceedings of the International Conference on Security and Management (SAM), p. 258. The Steering Committee of the World Congress in Computer Science, Computer Engineering and Applied Computing (WorldComp) (2016)
7. Majekodunmi, T.O., Idachaba, F.E.: A review of the fingerprint, speaker recognition, face recognition and iris recognition based biometric identification technologies (2011)

Comparative Analysis of Fingerprint Preprocessing Algorithms for Electronic Voting Processes

Kennedy Okokpujie(✉), Noma-Osaghae Etinosa, Samuel John, and Etta Joy

Department of Electrical and Information Engineering, College of Engineering, Covenant University, Ota, Ogun State, Nigeria
{kennedy.okokpujie, etinosa.noma-osaghae, samuel.john}@covenantuniversity.edu.ng, Joyetta96@gmail.com

Abstract. Fingerprints have been used for a long time as a very adequate method of distinguishing people. Matching fingerprints can be very difficult without a good algorithm for the scanners. Complexities especially during the verification phase can arise from problems with the algorithm scanners use. In elections and other forms of voting, security and proof of individuality remain some of the biggest obstacles to be surmounted. This paper proposes an ideal algorithm which could solve these challenges. The Zhan-Suen Thinning Algorithm and the Guo-Hall Parallel Thinning Algorithm were compared using standard parameters from the perspective of fingerprint technology and electoral requirements to deduce the better algorithm. MATLAB® was used for the comparative performance analysis of the two algorithms. The time taken to complete iterations, quality of output produced and reliability were the parameters used to carry out the comparative performance analysis of the two algorithms. The better algorithm, Guo-Hall's was recommended for implementation on e-voting systems. Guo-Hall's algorithm is very effective and has the capacity to curb electoral fraud when adopted in e-voting process.

Keywords: Thinning · Authentication · Enrollment · Verification

1 Introduction

Fingerprint enrollment is the sequences of steps taken to register a user's fingerprint. The user's fingerprint is extracted by a scanner, converted into a digital image, thinned and the minutia is extracted [3]. The minutiae are used to distinguish one fingerprint from the other. The minutia is used to form a template that is stored along with the user's demographic information in a database. Other important steps in authentication are verification and identification [5, 6]. Verification is the process where the biometric system compares a fingerprint to another which has been previously enrolled to

K.J. Kim et al. (eds.), *IT Convergence and Security 2017*,
Lecture Notes in Electrical Engineering 450,
DOI 10.1007/978-981-10-6454-8_27

determine if they are a match or got from the same finger (1:1 matching – as the system compares one query print to one enrolled print). Identification is when the biometric system compares a fingerprint of a user with the prints other enrolled users in the database to determine if that user is either known under a different name or a duplicate, false identity (1:N matching – N is the number of users which have been successfully enrolled) [1, 7]. Most times, a prospective user's inability to be enrolled or verified can be as a result of one or a combination of the following; age of the user, scarring and cuts on the finger, resolution and capture area of the sensor, biological, environmental and the nature of the user's work.

Since the introduction of biometrics in Nigeria, there have been many cases where users who have been previously enrolled are reported to have problems with authentication. One of such cases occurred during the March 2015 general elections where the Independent National Electoral Commission reported that there was a 41% failure rate in the use of biometrics (fingerprints) [2]. In the report, INEC cited various possible reasons for failure in the use of biometrics, but one solid reason is the presence of flattened or worn out fingerprints in users. In Nigeria for example, there are different databases managed by different private or governmental organizations, each with their own unique way of capturing users' fingerprint for data recognition. But an algorithm, selected through comparative analysis should form the foundation for developing uniform databases that make biometric enrollment and verification less cumbersome.

The aim of this paper is to compare with the aid of MATLAB® the efficiency of the Zhang-Suen thinning algorithm and the Guo-Hall parallel thinning algorithm, when implemented in an electronic voting machine that uses fingerprints scanner for enrolling voters. The Zhang-Suen thinning algorithm and the Guo-Hall parallel thinning algorithm were developed using Java. The efficiency each fingerprint algorithm and their comparative analysis were carried out using MATLAB®.

Fingerprints were acquired from a wide range of people from different walks of life. This was done to effectively prove that both algorithm works for different people. A fingerprint database that matches the fingerprints to each user's demographic information was created.

2 Background and Literature Review

2.1 Fingerprints in Electronic Voting

Enrolling is the initial step in the use of any biometrics. The image of the fingerprint is got by a sensor and a digital form of the image is made. The most essential feature of the fingerprint that differentiate one user from the other is extracted from the digital image, used to make a template and stored in a database with the unique user's information. To determine if the correct information was stored, a fresh sample template from the user is used to compare the stored sample for any match. Fingerprint samples got from an individual at different times are never exactly alike due to several factors among which are the way the user interacts with the biometric system and fingerprint changes due to injury and age. This is called Intra-class Variation. A threshold determines if two templates are a match. A score above the threshold is regarded as a match. In this work, a threshold of 0.40 which is the standard for iris recognition system was used. A detailed study of the intra-class and inter-class matching helps in determining False Rejection Rate and False Acceptance Rate.

When a template that is not stored in the database of the biometric recognition system is accepted, it is called False Acceptance and the frequency with which it occurs is called False Acceptance Rate. But when a template stored in the database of a biometric system is mistakenly rejected, it is called False Rejection and the frequency with which it occurs is called False Rejection Rate. These parameters are directly decided by the threshold value of the system. A high threshold value may lead to a higher false rejection rate while a low threshold value may lead to a higher false acceptance rate.

2.2 Zhang-Suen Fingerprint Thinning Algorithm

The Zhang-Suen algorithm is under a group of thinning algorithms which are known as iterative [3]. An iterative fingerprint algorithm is one which deletes pixels on the boundary of a pattern repeatedly until only a unit pixel-width thinned image remains. Thinning is done only on binarized images [4].

In the Zhang-Suen thinning algorithm, a digitized binary picture is defined by a matrix Z where each pixel Z (***i, j***) is either 1 or 0. The pattern consists of those pixels that have value 1. Each stroke in the pattern is more than one element thick. Iterative transformations are applied to matrix Z point by point according to the values of a small set of neighboring points [3, 4]. In this algorithm, an 8–point connectivity matrix is used which makes the iteration possible.

Zhang-Suen thinning algorithm matrix:

$$\begin{bmatrix} \boldsymbol{P9}, (\boldsymbol{i}-\mathbf{1},\boldsymbol{j}-\mathbf{1}) & \boldsymbol{P2}, (\boldsymbol{i}-\mathbf{1},\boldsymbol{j}) & \mathbf{3}, (\boldsymbol{i}-\mathbf{1},\boldsymbol{j}+\mathbf{1}) \\ \boldsymbol{P8}, (\boldsymbol{i},\boldsymbol{j}-\mathbf{1}) & \boldsymbol{P1}, (\boldsymbol{i},\boldsymbol{j}) & \boldsymbol{P4}, (\boldsymbol{i},\boldsymbol{j}+\mathbf{1}) \\ \boldsymbol{P7}, (\boldsymbol{i}+\mathbf{1},\boldsymbol{j}-\mathbf{1}) & \boldsymbol{P6}, (\boldsymbol{i}+\mathbf{1},\boldsymbol{j}) & P5, (\boldsymbol{i}+\mathbf{1},\boldsymbol{j}+\mathbf{1}) \end{bmatrix}$$

2.3 Guo-Hall Fingerprint Algorithm

This algorithm makes use of an 8-point connectivity matrix. Its iteration process is longer and takes more time to implement. The flowchart for the implementation Guo-Hall algorithm is shown in Fig. 1.

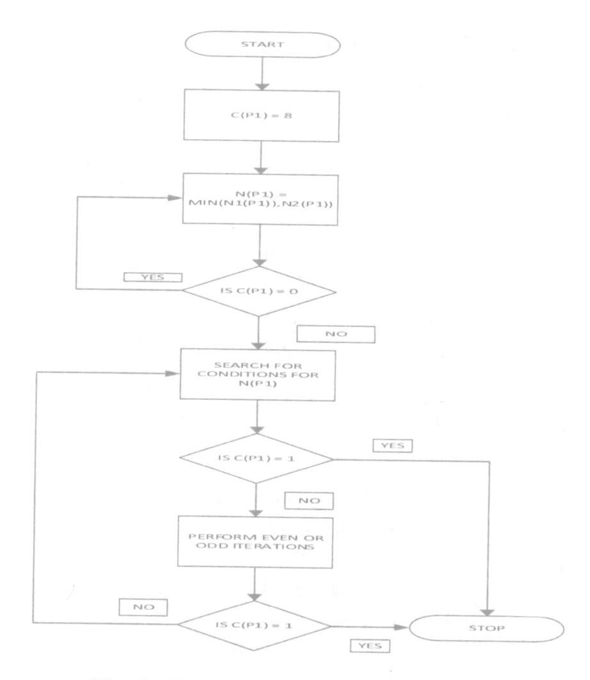

Fig. 1. Flow chart for Guo-Hall algorithm.

3 Methodology and Proposed Framework Data Flow

The comparative analysis was carried out using MATLAB®. The criteria used for comparison are:

1. Time taken to complete iterations (time taken to complete thinning process).
2. Quality of output produced (how close the fingerprint template is to the voter's fingerprint and the image's clarity).
3. Reliability (the algorithm's performance and behaviour over a defined period of time).

The FVC – 2000 data set provides the database from which thirty (30) fingerprints were randomly selected and used to carry out the comparative analysis.

4 Result and Discussion

(A) *Time taken to complete iterations:* the number of fingerprints used to carry out this test was increased in batches starting with just one fingerprint. The fingerprints on which each algorithm performed the thinning process were the same and in the same order. The time taken for both algorithms to complete the thinning process is tabulated below (Table 1).

Table 1. Fingerprint thinning time for each algorithm.

Batch	Prints number	Zhang-Suen	Guo-Hall
1	1	3.77 s	4.38 s
2	2	7.50 s	8.8 s
3	3	11.4 s	13.3 s
4	4	15.2 s	17.4 s
5	5	19.0 s	22.0 s
6	10	38.0 s	44.3 s
7	15	57.3 s	65.3 s
8	20	77.4 s	88.1 s
9	25	95.3 s	110.3 s
10	30	115.1 s	132.7 s

From the table above, it can be seen clearly that in comparison to Guo-Hall algorithm, Zhang-Suen algorithm performs its iterations (fingerprint thinning process) in a shorter time.

The graphical representation of the two algorithms' thinning time for each batch of fingerprint is shown in Fig. 2.

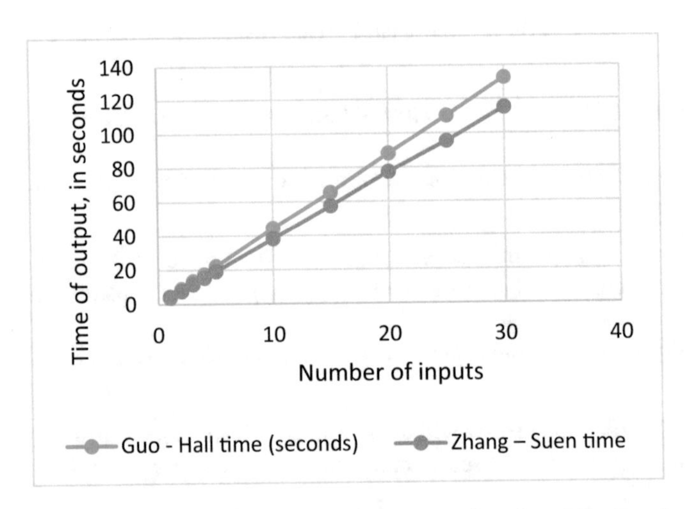

Fig. 2. Graphical representation of the time to complete iteration (thinning time) for both algorithms

This graphical representation also shows that Zhang-Suen fingerprint algorithm's time to complete iteration is shorter.

(B) *Quality of output produced*: the Guo-Hall parallel thinning algorithm produced a better output for twenty-eight out of thirty outputs, whereas the Zhang-Suen thinning algorithm produced only three comparatively better outputs for the same process. This is due to the fact that Zhang-Suen algorithm does not preserve the eight–point connectivity principle that is used to group and view images as one. Guo-Hall algorithm preserves this eight–point connectivity principle and thus produces a better output. A sample fingerprint input is shown in Fig. 3.

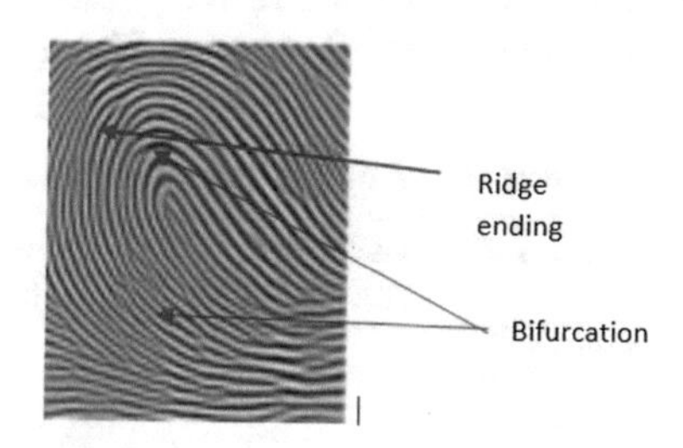

Fig. 3. Sample fingerprint input.

The iteration results (fingerprint templates showing their respective minutiae details) of the Zhang-Suen and Guo-Hall thinning algorithms details are on the sample fingerprint shown in Fig. 4.

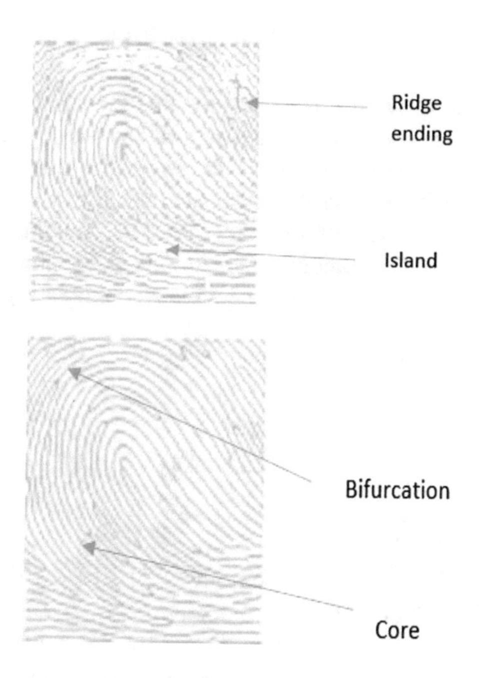

Fig. 4. Fingerprint template output of Zhang-Suen and Guo-Hall thinning algorithms respectively.

From the above images, it can be clearly seen that the Guo-Hall algorithm provides an output with better thinned lines than that of Zhang-Suen. Intricate lines like ridges, bifurcations and other form of measurable minutia are visible on the two outputs but they are less pronounced on the Zhang-Suen algorithm output.

(C) *Reliability*: the number of useful outputs each algorithm produced for each batch of fingerprints is summarized in the table below (Table 2):

Table 2. Comparison of useful outputs produced by each algorithm.

Batch	Prints number	Useful outputs (Zhang-Suen)	Useful outputs (Guo-Hall)
1	1	1	1
2	2	2	2
3	3	3	3
4	4	3	4
5	5	5	4
6	10	8	9
7	15	12	14
8	20	17	19
9	25	22	20
10	30	26	26

From the table above it was discovered that the number of useful fingerprint templates was quite consistent. The two algorithms showed an average non-useful output of 13%. The number of non-useful fingerprint template output climbed to its peak as the number of fingerprints in each batch to be thinned increased. The overall reliability test showed that the Guo-Hall thinning algorithm was more reliable especially as the number of fingerprints to be thinned increased.

The graphical representation of the table above is in Fig. 5:

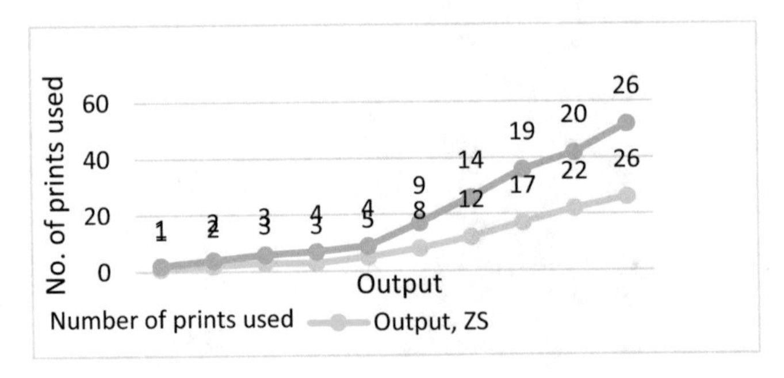

Fig. 5. Graph showing useful outputs for both Zhang-Suen and Guo-Hall algorithms

This graph shows that the Guo-Hall algorithm performs better in terms of reliability as it consistently produces more useful outputs as the number of fingerprints to be thinned increases.

5 Conclusion

The following can be deduced from the results of this comparative analysis:

1. The Zhang-Suen algorithm can be implemented in fingerprint enrollment and verification processes where timing is a serious concern.
2. The Guo-Hall algorithm should be implemented when the accuracy and reliability of fingerprints templates (output) are a major concern.
3. In all, despite the slower time for iteration shown by the Guo-Hall algorithm, it is superior to the Zhan-Suen algorithm in terms of output quality and reliability.

6 Future Work

The development of an hybrid algorithm that would incorporate the strengths of the Zhang-Suen and Guo-Hall algorithms.

Acknowledgement. This paper was sponsored by Covenant University, Ota, Ogun state, Nigeria.

References

1. Jain, A.K., Feng, J., Nandakumar, K.: Fingerprint matching. Computer **43**(2), 36–44 (2010)
2. Adebayo, H.: INEC says card reader test successful, admits 41% fingerprint failure [News], 10 March 2015. http://www.premiumtimesng.com/news/headlines/178264-inec-says-card-reader-test-successful-admits-41-fingerprints-verification-failure.html
3. Kocharyan, D.: A modified fingerprint image thinning algorithm. Am. J. Softw. Eng. Appl. **2**(1), 1–6 (2013)
4. Rosenfeld, A.: Image analysis and computer vision: 1988. Comput. Vis. Graph. Image Process. **46**(2), 196–250 (1989)
5. Daramola, S.A., Adefunmiyin, M.: Personal identification via hand feature extraction algorithm. Int. J. Appl. Eng. Res. **11**(7), 5148–5151 (2016)
6. Okokpujie, K., Olajide, F., John, S., Kennedy, C.G.: Implementation of the enhanced fingerprint authentication in the ATM system using ATmega128. In: Proceedings of the International Conference on Security and Management (SAM), The Steering Committee of the World Congress in Computer Science, Computer Engineering and Applied Computing (WorldComp), p. 258, January 2016
7. Daramola, S.A., Nwankwo, C.N.: Algorithm for fingerprint verification system. J. Emerg. Trends Eng. Appl. Sci. **2**(2), 355–359 (2011)

Design and Implementation of a Dynamic Re-encryption System Based on the Priority Scheduling

Duk Gun Yoon[1], Kyu-Seek Sohn[2(✉)], and Inwhee Joe[1]

[1] Hanyang University, Seoul, South Korea
{dgyoon, iwjoe}@hanyang.ac.kr
[2] Hanyang Cyber University, Seoul, South Korea
kssohn@hycu.ac.kr

Abstract. In this paper, a dynamic re-encryption system is proposed that prevents data from being decrypted and minimizes damage even if the encryption key is exposed by changing the encryption key according to the schedule and re-encrypting the stored data. We show that the proposed dynamic re-encryption system can reduce the probability from 100 to 0% that the RSA-512 ciphertext and the RSA-768 ciphertext are decrypted by the mathematical attack method over a given period. The proposed system can reduce the re-encryption time by 70% by implementing parallel processing with dynamic multithreading that checks the CPU usage of the server and adjust the number of threads actively. And the system ensures system reliability by preventing overflow and system interruption from occurring while maintaining server utilization at an average of 78%.

Keywords: Re-encryption · Priority scheduling · AES · RSA · Dynamic threading

1 Background and Needs

Sensitive data such as biometric data used in personal identity authentication of e-commerce such as FinTech [1], corporate secret data stored in cloud servers [2], and privacy data collected through IoT [3] are stored in encrypted form on the server. In order to enhance the security of the data, the encrypted data and its corresponding encryption key are stored in different servers, respectively [4, 5]. In order to increase the efficiency and consistency of data protection, data should be classified with the security level of the data according to the attribute of each data element [6].

In this paper, we propose a re-encryption scheduling system which can reinforce confidentiality and actively adapt to various server environment by re-encrypting the encrypted and stored data according to the schedule. The contents of this paper is composed of design and implementation of the proposed system, performance evaluation, and conclusion.

K.J. Kim et al. (eds.), *IT Convergence and Security 2017*,
Lecture Notes in Electrical Engineering 450,
DOI 10.1007/978-981-10-6454-8_28

2 Design of the Re-encryption System

2.1 System Architecture

The structure of the proposed re-encryption system consists of an encryption server, a key information database and a personal information database as shown in Fig. 1. Clients are the parties that send and receive data. They send and receive plaintext data through an encryption server. It is assumed that all components are connected to each other via secure networks such as VPN, TLS, and IPSec.

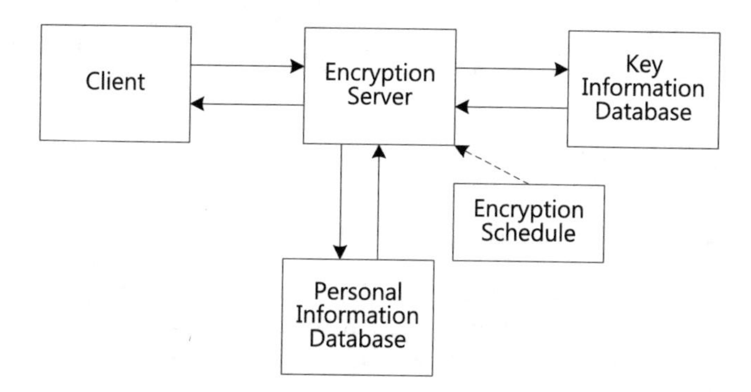

Fig. 1. Structure of the re-encryption system

When the encryption server (ES) receives a new data block from the client, it generates a random AES [7, 8] key and RSA [8] keys, encrypts the data with the AES key, and encrypts the AES key with the RSA public key, and then stores the encrypted AES key and the RSA public key in the key information database (KIDB). The ES stores the RSA private key and the encrypted data in the personal information database (PIDB).

To send data to the client, ES performs the decryption process in the reverse order of the encryption process. ES inquires the RSA private key and encrypted data from PIDB. ES inquires encrypted AES key and the RSA public key from KIDB. ES decrypts the AES key with the RSA private key and decrypts the ciphertext of data with the AES key.

ES generates a new AES key and new RSA keys randomly according to the re-encryption schedule and performs the re-encryption process. At first, ES performs the decryption process and then performs the encryption process with the AES key and RSA keys newly generated.

The re-encryption system can reduce the probability that they are exposed at the same time by storing the encryption key and the encrypted object in different databases, respectively. If the encryption key has been re-encrypted even though it is exposed, the new encrypted data can not be decrypted with the old encryption key as shown in Fig. 2. The re-encryption system not only enhances the security of data and encryption keys by providing forward secrecy, but also ensures human security by eliminating the need for administrator intervention in the re-encryption process.

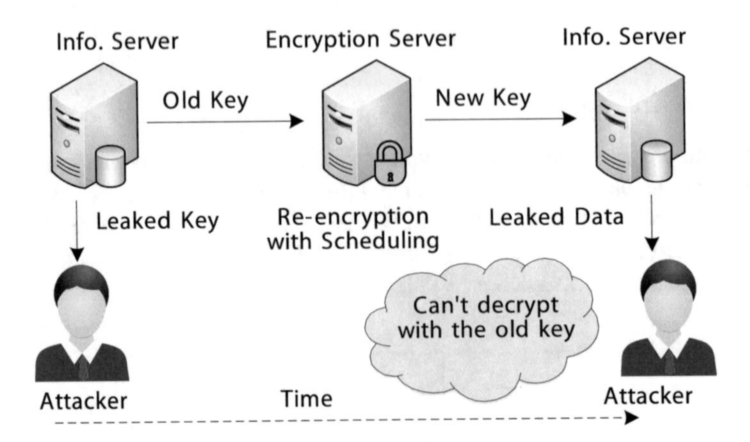

Fig. 2. Security against exposure of the encryption key and ciphertext

2.2 Priority Scheduling for Re-encryption

Table 1 lists the external factors used to determine the re-encryption priority to be given to each data block when re-encryption scheduling. Depending on the field or context to which the re-encryption system is applied, the external factors can be set differently.

Table 1. External factors used in determination of re-encryption priority.

Criteria	Description
Security level	Integer value from 1 to 5 (larger means higher) [6]
Schedule cycle	Scheduling period in terms of days, hours, minutes, seconds, etc.
Need of schedule	This factor describes whether the data should be re-encrypted
Schedule date	The last re-encryption time
Use of data	This factor represents the data is being used or not
Data limited	The largest number of data blocks without overflow

The priority of the data is determined according to the security level of the data. The higher the security level the higher priority is. However, priority scheduling requires priority reordering because starvation may occur if data having a higher priority than the waiting data continues to be received [9]. In the re-encryption system, we solved the starvation problem by increasing the priority of the data that has not been re-encrypted in its re-encryption period. Figure 3 is a pseudo code list of the procedure that takes precedence in considering the external factors given in Table 1. The parameter alpha has an arbitrary integer value of 1 or more. The higher the value of alpha, the slower the rate of increase in priority.

```
function setPriority()
begin
    if needSched == "Y" and usedData == "N" then
        Age = floor((currentDate - ScheduleDate - 0.1)
                    / (SCHEDULECYCLE * alpha))
        setPriority = min(HIGHESTPRIORITY,
                          SecurityLevel + Age)
    end if
end
```

Fig. 3. A pseudo code list for determining prioity of a data block

2.3 Implementation of the Re-encryption System

The re-encryption system was implemented in Java, an object-oriented language. The class diagram of the re-encryption system is shown in Fig. 4. Re-encryption scheduling is implemented in ScheduleEnc class.

Random key generation and encryption and decryption for re-encryption are implemented by API provided by Java. The EncAES class and the DecAES class are constructed based on the Cipher class, KeyGenerator class, and the SecureRandome class. The EncRSA class and the DecRSA class are constructed based on the Cipher class and the KeyPair Generator class. Objects of all these classes are processed through the Schedule Thread class which is a thread for re-encryption.

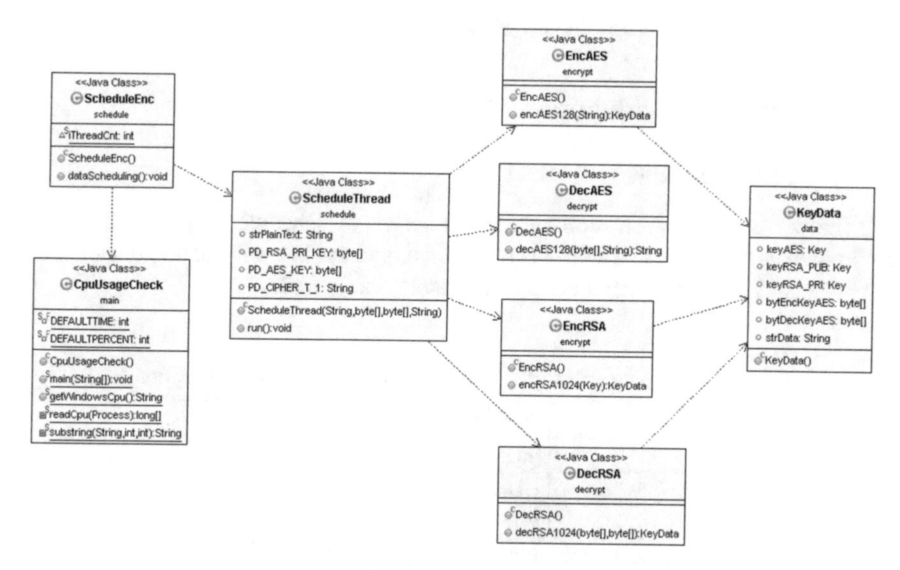

Fig. 4. Class diagram of the re-encryption system

2.4 Dynamic Multithreading

We have constructed a multi-threaded parallel processing system to speed up re-encryption of the re-encryption scheduling system [9]. In order to solve the trade-off problem of re-encryption operation speed and system stability, the re-encryption scheduling system actively determined the number of proper threads according to the CPU usage of the server. Figure 5 is the pseudo code list of an algorithm that determines an appropriate number of threads. We adjusted the number of threads to keep the CPU utilization between min cpu usage and max cpu usage. As an example, the values of min cpu usage and max cpu usage are 60% and 80%, respectively. The range of CPU utilization can be changed according to the operating conditions of the system.

```
function paralleled_re-encryption ()
begin
    while not empty_data_queue
        if CpuUsage < MIN_CPU_USAGE
            create a thread for paralleled_re-encryption
        else if CpuUsage > MAX_CPU_USAGE
            cancel a thread
        end if
        do re-encryption
    end while
end
```

Fig. 5. A pseudo code list of the dynamic re-encryption scheduling

3 Performance Evaluation

3.1 Test Environments

For the performance evaluation of the re-encryption scheduling system, the server environments as shown in Table 2 is configured, and the key information database and the personal information database are separately configured.

Table 2. External factors used in determination of re-encryption priority.

Item	Specification
CPU	Intel Core 2 Duo, 2.53 GHz
RAM	DDR3, 4 GB
OS	Windows7 64 bit
Programming tools	JDK 1.8, Eclipse Mars, MS SQL 2005

3.2 Security

The re-encryption system uses RSA and AES algorithms. To decrypt the ciphertext of the data, the AES key must first be decrypted with the RSA private key. Attack methods for RSA algorithms that can be used when an attacker does not know the RSA private key are brute force attacks, mathematical attacks, timing attacks, and chosen ciphertext attacks [8]. This performance evaluation is a comparative evaluation of security performance against mathematical attacks. The mathematical attack algorithm is shown in the following equation [10, 11]:

$$\exp\left((c+o(1))(\log n)^{\frac{1}{3}}(\log\log n)^{\frac{2}{3}}\right) \tag{1}$$

This algorithm is the best attack algorithm for RSA encryption among the known mathematical attacks, and based on this, RSA-512 was decrypted in 1999 [10], and RSA-768 was decrypted in 2 years in 2009 [11]. At present, it is anticipated to shorten the time required due to the development of CPU performance, but this data is compared for evaluation.

It is assumed that there is no possibility that the encrypted AES key of the KIDB and the encrypted data of the PIDB will leak together during one re-encryption scheduling period. This assumption can be realized by storing both fake data and a pseudo symmetric key in each database.

Let' s say the probability of deciphering a six-month mathematical attack on data encrypted with the RSA-512 algorithm is 100%. If the scheduling period of the re-encryption scheduling system that encrypts the AES cryptographic key with the RSA-512 algorithm is set shorter than 6 months and the re-encryption is continuously performed, the decryption possibility can be reduced from 100% to 0.

3.3 Re-encryption Speed

The time spent in the re-encryption operation is the sum of the time took to perform priority calculation and assignment, decryption, and encryption. When the amount of data is 10,000 blocks, it takes about 24 min to re-encrypt and the average CPU usage is 34%. When the re-encryption scheduling system is composed of 1, 2, 3, and 4 threads, the relationship between the amount of data and the re-encryption operation time is shown in Fig. 6.

When the number of threads increases from 1 to 2, the re-encryption time decreases by half, but when the number of threads increases to 3 or 4, the reduction rate of re-encryption time decreases. Experimental results show that the minimum re-encryption time for four threads is reduced to 30% for single threads. That is, the re-encryption time is reduced by about 70%.

Average CPU utilization was around 34% when processing re-encryption with a single thread. The result of evaluation shows that when the number of threads increases as 1, 2, 3, and 4, the average CPU usage rate is also increased to 34%, 55%, 68%, and 78%, respectively. When the number of threads is 5 or more, the CPU utilization rate becomes 100% and the system crashes due to memory or stack overflows. Figure 7

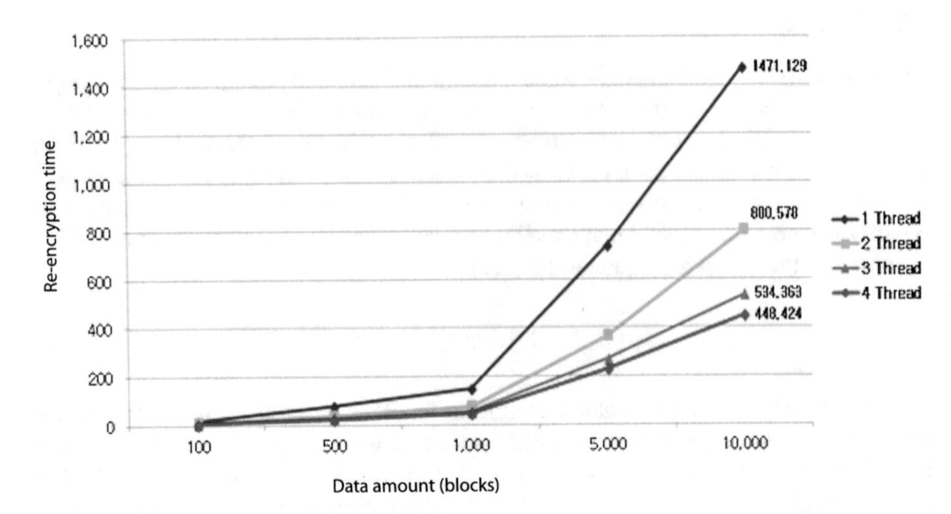

Fig. 6. Data amount vs. re-encryption time with the number of threads as a parameter (Re-encryption time includes priority calculation time, decryption time, and encryption time)

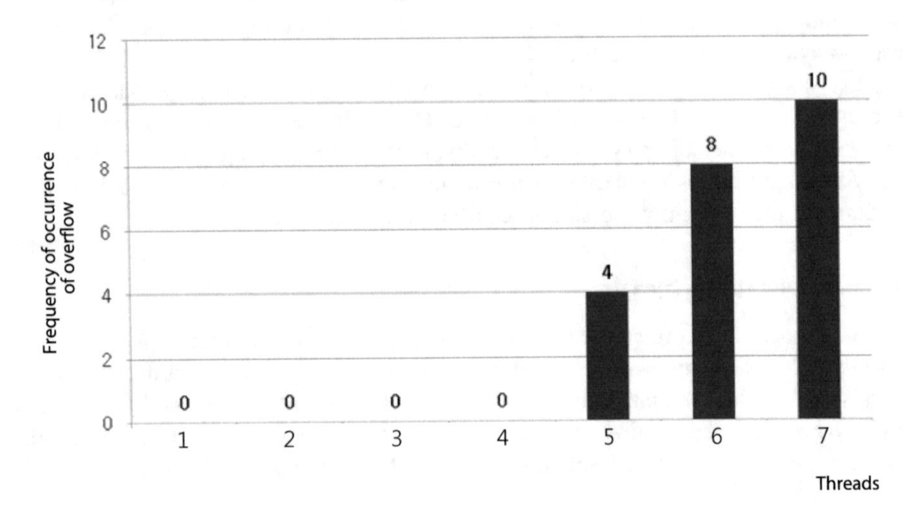

Fig. 7. D The number of threads vs. overflow frequency

shows the frequency of overflow occurrences by performing 10 times of re-encryptions of each 10,000 data blocks for each case of the number of threads.

Any overflow did not occur when there were fewer than four threads. Therefore, when four threads are used, the re-encryption time is the shortest and the CPU utilization is the highest.

4 Conclusion

The proposed re-encryption scheduling system sets the re-encryption priority of each data at a period of the schedule and selects re-encryption target data according to the priority. The system decrypts the selected data and encrypts it by applying the newly generated encryption key. The performance evaluation showed that if the AES key of the re-encryption system is encrypted with RSA-512 or RSA-718 and the re-encryption period is less than 6 months or less than 2 years, respectively, decryptability of the encrypted AES key can be reduced to 0. The re-encryption system can have the appropriate number of threads by which the system can achieve maximum re-encryption speed and CPU usage without overflow. In relation to this, the performance evaluation showed that the optimal number of thread is 4, reduction ratio of the re-encryption time is 70%, and the maximum average CPU usage is 78% while the system crash frequency can be kept at zero.

As a further study, we will search a way to apply the re-encryption system to faster and more energy efficient secure data storages and crypto key distribution mechanisms.

References

1. Roberts, J.J.: How Biometrics are Worse than Passwords. Fortune, 12 May 2016. http://fortune.com/2016/05/12/biometrics-passwords/
2. Shastri, A., Sharma, P.: Data vault: a security model for preventing data theft in corporate. In: ICTCS 2016, vol. 142. ACM (2016)
3. Botta, A., et al.: Integration of cloud computing and internet of things: a survey. Future Gener. Comput. Syst. **56**, 684–700 (2016)
4. Smith, A.D.: Maintaining secrecy when information leakage is unavoidable. Thesis of Ph.D. in Computer Science, MIT (2004)
5. Kohler, J., et al.: An approach for a security and privacy-aware cloud-based storage of data in the semantic web. In: ICCCI, pp. 241–247. IEEE (2016)
6. Shaikh, R., SasiKumar, M.: Data classification for achieving Security in cloud computing. Proc. Comput. Sci. **45**(1), 493–498 (2015). Elesevier B.V.
7. NIST, FIPS PUB. 197: Advanced Encryption Standard (AES), November 2001
8. Stallings, W.: Cryptography and Network Security, 6th edn. Pearson Education India, New Delhi (2013)
9. Galvin, P.B., Gagne, G., Silberschatz, A.: Operating System Concepts. Wiley, Westminster (2013)
10. Cavallar, S., et al.: Factorization of a 512-bit RSA modulus. In: Advances in Cryptology — EUROCRYPT 2000, pp. 1–18. Springer, Heidelberg (2000)
11. Kleinjung, T., et al.: Factorization of a 768-bit RSA modulus. In: Advances in Cryptology – CRYPTO 2010, pp. 333–350. Springer, Heidelberg (2010)

Network Forensics Investigation for Botnet Attack

Irwan Sembiring(✉) and Yonathan Satrio Nugroho

Satya Wacana Christian University, Diponegoro 52-60, Salatiga, Indonesia
irwan@staff.uksw.edu, yonathansatrio@gmail.com

Abstract. Nowadays the internet users manipulated by several web applications which instruct them to download and install programs in order to interfere the computer system stabilities or other aims. Most users didn't realize that the applications might have been added with some malicious software such as Worms, and Trojan horse. After the malware infected the victim's computer, they made the machine to conduct for to the master's purposes. This process known as botnet. Botnet is categorized as difficult detected malware even with up-to-date antivirus software and causing lot of problems. Network security researcher has developed various methods to detect Botnet invasion. One of the method is forensics method. Network forensics is a branch of Digital forensics which the main task is to analyze the problem (e.g. Botnet's attack) by identify, classify the networks traffic and also recognize the attacker's behavior in the network. The output of this system will produce the pattern recognition of Botnet's attack and payload identification according to Network Forensics Analysis.

Keywords: Malware · Botnet · Network forensics

1 Introduction

The main aim of botnets or robot networks creation were to help the network administrator to operate the channels of Internet Relay Chat (IRC) services. But, then some people also paying attention about the bot inside botnet which give the instruction to the machine, as they seen it might be the weakness and could be functioned as criminality method. Attackers simply identify botnet is working under the administrator (master), fully-controlled via dedicated server called Command & Control Server (C&C Server). Botnet is working through the computers that have been infected by malicious software (malware). The malicious software tends to act as Trojan horse, worms, and backdoor [1]. Most of the internet users might have seen the popular case such as redirect link/hijack link which turned into spamming, and the malicious executable software which is Trojan or worms after the installation (malicious executable) software. Zeus botnet is indicated as one of the most dangerous C&C botnet because of its capability in stealing the online account identity such as username and password. It generally use to steal bank accounts for money profits and steal any account typed in user computer that infected by Zeus bot such as email accounts, or social websites accounts. At least in 2009, Zeus botnet known as the largest widespread malware,

K.J. Kim et al. (eds.), *IT Convergence and Security 2017*,
Lecture Notes in Electrical Engineering 450,
DOI 10.1007/978-981-10-6454-8_29

infected in over 200 countries and affecting approximately 75,000 computers. Spreaded to victims by drive download or by email spamming [2].

Network forensic analysis takes the important role to overcome the various things of cybercrime such as identity theft, fraud, and child pornography. Network forensic analysis is a branch of Digital Forensics. The tasks are analyzing and classifying the network traffic, attack behaviors, packet, and system log based on cases being happened. Based on problem correlate to botnets, this paper is expected to contribute in how to recognize and identify the pattern of C&C Botnets (Zeus Botnets) attack according to Network Forensic Analysis. Classifying the C&C Botnet attack pattern based on Network forensic analysis method, visualizing and investigate in order to result the rule that could be applieds on Intrusion Detection System (IDS) in further research. In this paper, botmaster is installing Zeus botnet builder and Cpanel Server on virtual machine with Windows Seven operating system. Victims PC are also created on virtual machine with Windows XP operating system. For network forensics analysis requirements, there are several application to use; NetWitness Investigator, Wireshark, Fiddler2, and HxD (Hex Editor).

2 Related Works

The previous related works used as references are; "Analysis and Detection of the Zeus Botnet Crimeware". This study was a research title from M.I. Laheeb et al. in 2015, presented a design and implement a Host Botnet Detection Software (HBD's) to the victims PC to detect the signature and pattern of Zeus botnet attack. Then they analyzed the botnet activity and made removal of Zeus botnet by using Ollydbg reverse engineering tool and penetration operation [3]. The next work is "Peer-to-Peer Botnets: Overview and Case". This study was a research title from J.B. Grizzard et al. in 2007, where they present an overview of peer-to-peer botnets. Their case study of the Trojan. Peacomm.bot demonstrated one implementation of peer-to-peer functionality used by a botnet [4]. Another related work used as reference is "Network Forensic Analysis Using Growing Hierarchical SOM" presented in 2013 by S.Y. Huang et al. SOM (self-organizing map) is a data mining algorithm, used as a clustering and classification method. The research compared GHSOM (Growing Hierarchical SOM) with K-Means method to conclude the pattern of botnet attack from sample dataset of log network traffic data had been collected. After the comparison, it was concluded that GHSOM had a better value of botnet attack's recognition accuracy than K-Means Clustering. The analyze result of GHSOM could be used as a reference in application of digital forensics [5]. Zeus botnet is one of C&C botnet, using bot as a backdoor, or establishing the access to the victims PC system [6]. The bot infections are widespread automatically, directly directed by the attacker using worm to find the potential space of networks or subnet. Another way of botnet attack is by using some fake web applications, which have been added by Trojan. The internet users might be influenced by the instruction that instruct them to download and install the application to their computer. They didn't realize that they also installing bot to their computer after the installation of fake software downloaded on the internet. Network forensics analysis defines as a medium to apply or implement forensics science to the computer networks in order to discover the source of networks crimes. The objective is to identify malicious activities from the traffic logs, discover their details, and to assess the damage [7]. Network forensics is basically about monitoring, capturing, analyzing the network traffic and investigating the security policy violations. Forensic specialist monitors the network continuously and stores a copy of all or the relevant packets, depending upon the policy, in a prescribed format for future analysis. If any attack is found, the type of attack is determined and the source of the attack is investigated. Forensic specialists can attribute the attacker by proper monitoring, capturing, and analysis of the network traffic and by proper investigation. There are six steps of the general process of network forensics [8]: (1) Preparation and Authorization, (2) Collection of Network Traces, (3) Preservation and Protection, (4) Examination and Analysis, (5) Investigation and Attribution, dan (6) Presentation and Review (see Fig. 1)

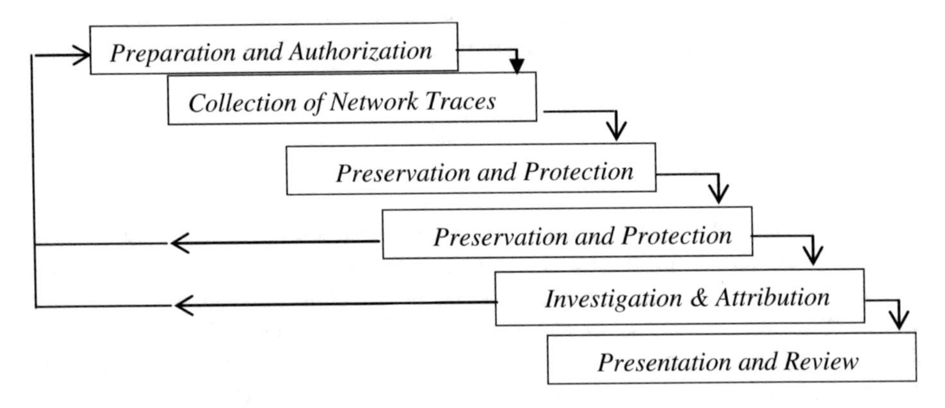

Fig. 1. The generic framework for network forensics

Figure 1 shows the generic framework for network forensics where it can be a guidelines to help the forensic specialists in determine their works phase. The first step is preparation and authorization. Preparation is step of deploying the requirement tools such as intrusion detection system, packet analyzers, and firewalls. These tools need to be configured and deployed at various strategic points on the network. The required authorizations to monitor the network traffic are obtained so that privacy of individuals and the organization is not violated. Any unauthorized events and anomalies noticed will be analyzed. The presence of the attack is determined from various parameters. The second step is Collection of Network Traces. Collection is the most difficult part as traffic data changes at rapid pace and it is not possible to generate the same trace at a later time. The amount of data logged will be enormous requiring huge memory space capacity. The original data obtained in the form of traces and logs is stored on a back up device. A hash of all data is taken and the data is protected. Another copy of the data will be used for analysis and the original collected network traffic is preserved. The collected data is classified and clustered into groups so that the volume of data to be stored may be reduced to manageable chunks. The evidence is searched methodically to extract specific indicators of the crime. The indicators are classified and correlated to deduce important observation using the existing attack patterns. The information obtained from the evidence traces is used to identify who, what, where, when, how and why of the incident. The observations are presented in an understandable language to organizations management and legal personnel while providing explanation of the various standard procedures using used to arrive the conclusion. The results are documented to influence future investigation and in improvement of security products.

3 Network Forensics System Architecture

This paper is using the design of Network Forensics System (NFS) to discover the source of attack, as shown in Fig. 2 [8], consist of: (1) Capture Module, (2) Analysis Module, and (3) Presentation Module (see Fig. 2).

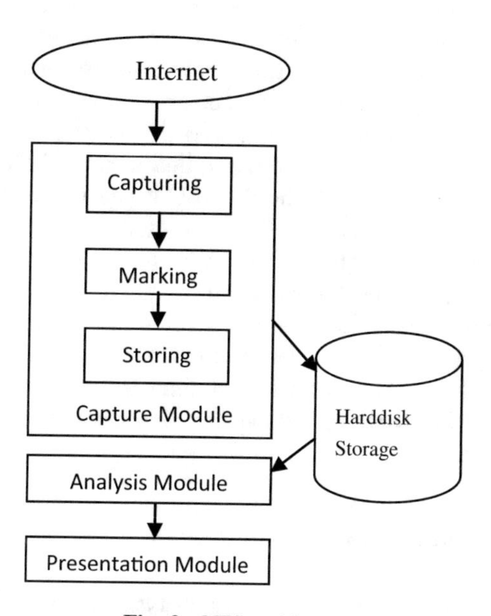

Fig. 2. NFS architecture

The proposed model (Network Forensics System Architecture) is used to identify the patterns, behaviors, and the characteristics of Zeus Botnet while attack the victims PC. The system is designed to capture the network traffic from the network interface of the host, mark the relevant traffic packets, analyze the marked packet and display valuable information. The capture module consists of three sub modules: capturing, marking, dan storing process. The capturing process captures all packets that passing through the network interface of host PC by using Wireshark. This stage is the first process of identifying the malicious activities inside the network traffic catched by Wireshark. The next step is the marking process. It marks the data of the packet capture (pcap) according to the researcher requirement. This paper researched the Zeus Botnet while passing through HTTP protocol only, with time interval of the attacker's server being communicated with victims PC as a parameter. The HTTP choosed because Zeus Botnet usually communicates via HTTP GET/POST requests to a specific server resource (Uniform Resource Identifier). In the marking process, there is a reduction process of information and only the relevant or required data to gained and analyzed. The storing process defines as saving the marked packets into *hard disk* of the *host system for further analysis*. This process ensure that the only the actual packets used for Zeus Botnet attack are stored, so it has the less or smaller size than the first packet captured. The second module is the Analysis Module. The analysis module analyzes the log file stored in the host system by the capture module, in order to discover the source of network attack. This paper analyzes the packet capture (pcap) from six computers which have been infected by Zeus botnet. There are some information to be analyzed correlated with the infected machine while accessing the internet [9]: (1) It is abnormal for most users to connect to a specific server resource repeatedly and at periodic intervals. There might be dynamic web pages that periodically refresh content, but these legitimate behaviors can be detected by looking the server responses; (2) The first connection to any web server will always have response greater than 1 kB because these are web pages. A response size of just 100 or 200 bytes is hard to imagine under usual conditions; (3) Legitimate web pages will always have embedded images, JavaScript, tags, link to several other domains, link to several file paths on the same domain, etc.; (4) Browser will send the full HTTP headers in the request unless it comes from a man-in-the-middle attack. The third module is the Presentation Module. It presents the result output from the analyze process of six computers infected by Zeus botnet. Those six victims are one-by-one identified and investigated to find the general pattern of Zeus botnet attack. The next processes are clustering and classify the characteristics conclusion of Zeus botnet from comparing of the threshold values. The presentation also includes the conclusion of time interval and sessions while the attacker communicate with victims, and mention what they got after infect the victims beside the user account.

4 Results and Discussion

The System made based on network forensics analysis with some certainty as follow: six computers infected by Zeus Botnet are using Windows XP operating system, while the attacker (botmaster) is using Windows Seven operating system. The attacker's pc name is Windows 7 with 192.168.0.3 address, and the rest of computers are victims. Each of victim computers is analyzed and investigated according to Network Forensics System design. The first step is capturing the packets from network traffic by using Wireshark. The capturing process is followed by the activity of accessing the internet by the every victim's computer simultaneously as long as one hour. After the packet capture (pcap) of each victim's computer were collected, then the next step is marking process. The marking process decides which is the relevant requirement needed by forensics specialist to investigate the ongoing case. In this research, HTTP protocol was choosed as the Zeus Botnet which usually communicates via HTTP GET/POST to the Command & Control Server owned by the attacker. The marking process begins with eliminate the protocols catched by Wireshark except HTTP protocol. After gaining and marking the packet capture (HTTP protocol) from every victim's computers, there is some decrement amount of pcap size, as it shown in the Table 1.

Table 1. Marking process

PC name	Before marking (kB)	After marking (kB)
Xp	3,138	358
pc2	513	196
pc3	1,107	252
pc4	2,013	395
pc5	6,272	330
pc6	1,106	389

As shown in Table 1, there is significant decrement of the packet where it was used for further analysis. The decrement size of the packets after marking process also make some benefit to the forensics specialist because its size become not too large and could save some capacity of storage. The following step after marking process is the analysis module. The analysis module analyzes the packet captured from the marking process. In this case, the pcap to analyze is pcap with HTTP protocol. The analysis module is driven to every victim's computers in order to gain the characteristics of Zeus Botnet. Here is the analysis process of XP computer, as one of six total victims captured by Wireshark (see Fig. 3). Figure 3 is one example of six pcap catched by wireshark. Almost all of the victim computers have the same characteristic when accessing the internet. XP computer, which has 192.168.05 IP address often communicate with specific address of server (192.168.0.3). From 20 dataset taken, there is only one unique URI connected and make repetitively 10 times of communication. One unique URI connected is less than the threshold, while one unique URI connected and make repetitively 10 times of communication is counted greater than the threshold value. The analysis of pcap also uses NetWitness Investigator to make the investigation easier. By using NetWitness Investigator, the other pattern or characteristics of Zeus Botnet could be identified.

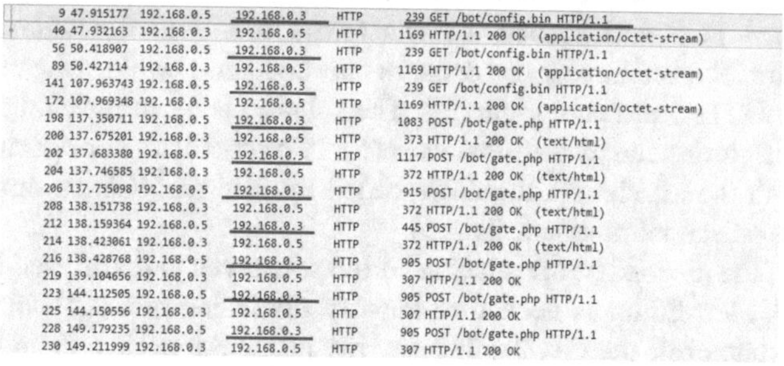

```
  9 47.915177  192.168.0.5  192.168.0.3  HTTP   239 GET /bot/config.bin HTTP/1.1
 40 47.932163  192.168.0.3  192.168.0.5  HTTP  1169 HTTP/1.1 200 OK  (application/octet-stream)
 56 50.418907  192.168.0.5  192.168.0.3  HTTP   239 GET /bot/config.bin HTTP/1.1
 89 50.427114  192.168.0.3  192.168.0.5  HTTP  1169 HTTP/1.1 200 OK  (application/octet-stream)
141 107.963743 192.168.0.5  192.168.0.3  HTTP   239 GET /bot/config.bin HTTP/1.1
172 107.969348 192.168.0.3  192.168.0.5  HTTP  1169 HTTP/1.1 200 OK  (application/octet-stream)
198 137.350711 192.168.0.5  192.168.0.3  HTTP  1083 POST /bot/gate.php HTTP/1.1
200 137.675201 192.168.0.3  192.168.0.5  HTTP   373 HTTP/1.1 200 OK  (text/html)
202 137.683380 192.168.0.5  192.168.0.3  HTTP  1117 POST /bot/gate.php HTTP/1.1
204 137.746585 192.168.0.3  192.168.0.5  HTTP   372 HTTP/1.1 200 OK  (text/html)
206 137.755098 192.168.0.5  192.168.0.3  HTTP   915 POST /bot/gate.php HTTP/1.1
208 138.151738 192.168.0.3  192.168.0.5  HTTP   372 HTTP/1.1 200 OK  (text/html)
212 138.159364 192.168.0.5  192.168.0.3  HTTP   445 POST /bot/gate.php HTTP/1.1
214 138.423061 192.168.0.3  192.168.0.5  HTTP   372 HTTP/1.1 200 OK  (text/html)
216 138.428768 192.168.0.5  192.168.0.3  HTTP   905 POST /bot/gate.php HTTP/1.1
219 139.104656 192.168.0.3  192.168.0.5  HTTP   307 HTTP/1.1 200 OK
223 144.112505 192.168.0.5  192.168.0.3  HTTP   905 POST /bot/gate.php HTTP/1.1
225 144.150556 192.168.0.3  192.168.0.5  HTTP   307 HTTP/1.1 200 OK
228 149.179235 192.168.0.5  192.168.0.3  HTTP   905 POST /bot/gate.php HTTP/1.1
230 149.211999 192.168.0.3  192.168.0.5  HTTP   307 HTTP/1.1 200 OK
```

Fig. 3. XP computer in analysis module

The attacker's server ensures the access to the victim computers by giving the script or instruction (config.bin) to the bot inside the infected machine. Other way to conclude the time session communication between the attacker and the victims is by giving the graph pattern. The graph is created using NetWitness Investigator and defines as general pattern of the whole victims investigation (see Fig. 4).

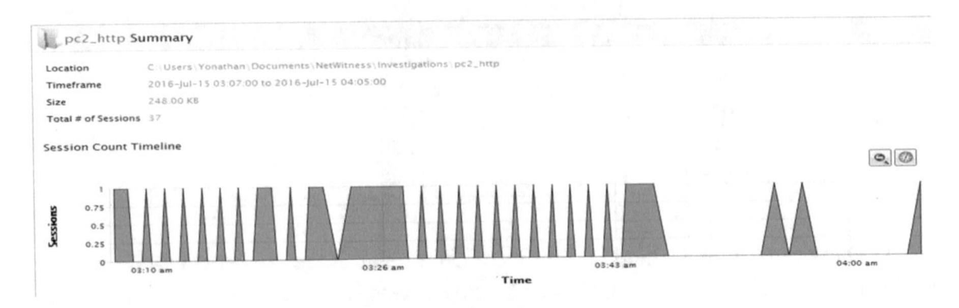

Fig. 4. Interval time (session)

Figure 4 shows the time interval or session communication between the attacker and PC2 for one hour. The pattern is repetitively constant as the attacker trying to establish the access to the victim computer in every 20 or 21 s. The other victim computers have also been investigated, and the result was not too significantly different with the Fig. 4. The next step is analyzing the payload that has been captured in the capturing module. *Payload* is the encrypted data and formed as hexadecimal numbers. The contents of payload can be consist of various file, such as document, picture, *audio*, or other kinds, separated into fragments and encrypted with its hash algorithm. The forensics specialist needs to decrypt and extract the content inside payload to find out the attached file. The purpose of payload analysis is to discover files gained by Zeus Botnet when they infect the machine and steal the internet account. It is also needed as the requirement of network forensics analysis to find out all important things under investigations. The simulation of payload analysis begins with PC6 download the attachment file from gmail account. The network where PC6 is located has identified as the infected networks. All activities were catched by network traffic capturing sensor, including when PC6 downloaded the attachment file. As information, gmail uses HTTPS (Secure Socket Layer) protocol for the operation, while wireshark does not recognize the HTTPS protocol traffic [9]. The solution is by using Fiddler2 to replace wireshark in capturing the packets that use HTTPS protocol in their operation. Fiddler2 is able to catch the attachment file downloaded by PC6 and able to show the packet payload in hexadecimal numbers (Fig. 5).

After that, the payload of the attachment file was saved and analyzed by using Hex Editor (HxD). Hex Editor is needed to identify file inside hexadecimal number, and also as the extractor of the file. To find any file inside the packet, there are two main characteristic to be noticed called File Signature. File signature consist of Header Offset and Trailer Offset. Header offset is an initial byte series of the payload, and Trailer

Fiddler Web Debugger
Composer
Log
Filters
Timeline
Statistics
Inspectors
AutoResponder
Request Headers
Client
Accept: */*
Accept-Encoding: gzip, deflate
Accept-Language: en-us
Miscellaneous
Transport
Connection: Keep-Alive
Readonly

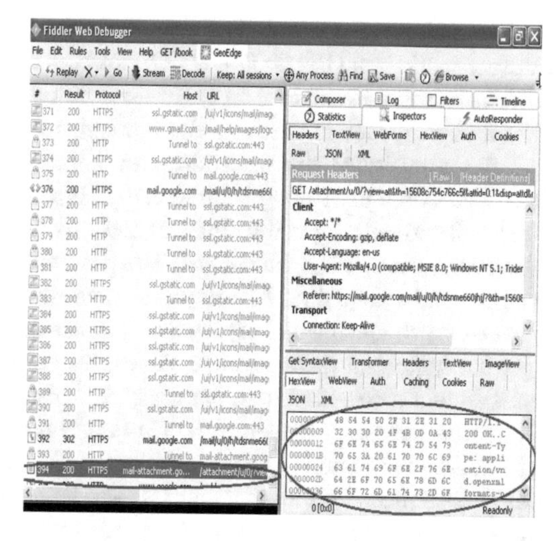

Fig. 5. The attachment file and payload catched by Fiddler2

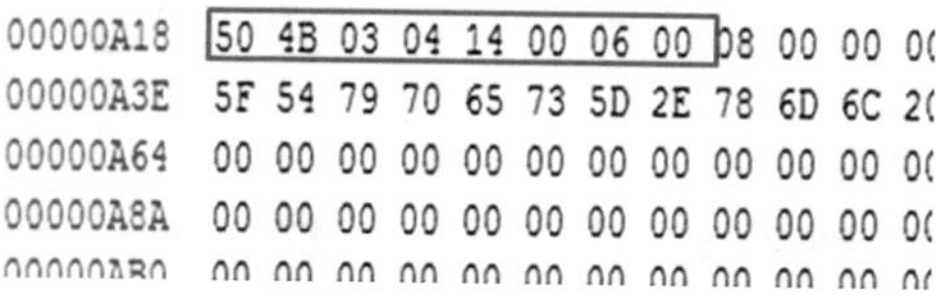

00000A18 50 4B 03 04 14 00 06 00 08 00 00
00000A3E 5F 54 79 70 65 73 5D 2E 78 6D 6C
00000A64 00 00 00 00 00 00 00 00 00 00 00
00000A8A 00 00 00 00 00 00 00 00 00 00 00

Fig. 6. The header offset

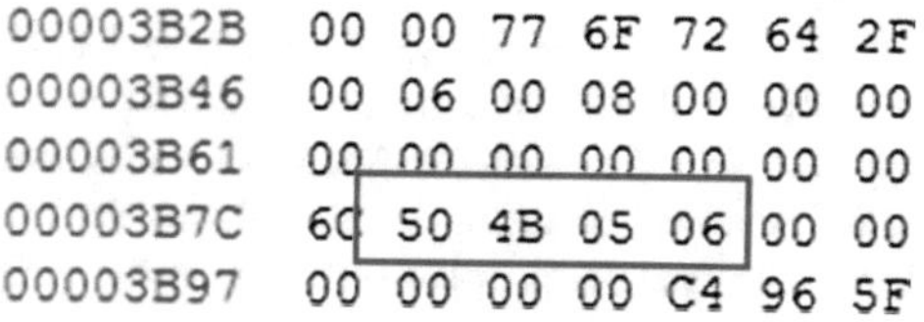

00003B2B 00 00 77 6F 72 64 2F
00003B46 00 06 00 08 00 00 00
00003B61 00 00 00 00 00 00 00
00003B7C 6C 50 4B 05 06 00 00
00003B97 00 00 00 00 C4 96 5F

Fig. 7. The trailer offset

offset is a unique series of byte located at the end of payload. Packet has been analyzed and there is File signature as it indicates the payload. The Header offset is shown in Fig. 6, and the Trailer offset is shown in Fig. 7.

5 Conclusion and Future Work

In summary, using the Networks Forensics System in analyzing Botnet attack is very useful to detect the characteristics, patterns and behaviors of the attack. The proposed architecture of network forensics system is proved to find out the conclusion of

Zeus Botnet based on certainty parameter. The conclusions described as follows: (1) Generally Zeus Botnet works via GET/POST HTTP protocol requests to a specific server resource (Uniform Resource Identifier); (2) The indication of the Zeus Botnet attack can be seen from several factors, such as the amount number appearances of unique domain (URI), the intensity of specific address to communicate with, and the time interval (session) when communicating with specific address; (3) The Network forensic analysis is basically identic by using forensic *tools to help the investigation such as capturing, clustering, classifying and etc.*; (4) The presentation is a stage which make the results become understandable. In this case, presentation of Zeus botnet attack by graph is make easier to understand; (5) By using forensics method, the packet can be analyzed to get the important payload information; further works described as follows: (1) The identified and detected pattern can be used and implemented directly to the server as the Intrusion Detection System rule; (2) Future research can be driven with other protocol to decide the pattern of Zeus Botnet attack; (3) The current proposed architecture could be replaced with other forensic model to prove validation in recognizing attack, otherwise the attack is replaced by another attack but still using the current proposed Network Forensic System model; (4) Use another method to extract the file from the payload.

References

1. Shaikh, A.: Botnet Analysis And Detection System. School of Computing, Napier University (2010)
2. Trend Micro Inc.: Zeus: A Persistent Criminal Enterprise. Trend Micro, Incorporated Threat Research Team (2010)
3. Laheeb Mohammed Ibrahim: Analysis and Detection of the Zeus Botnet Crimeware. Mosul University, Mosul (2015)
4. Grizzard, J., Sharma, V.: Peer-to-peer botnets: overview and case study. In: HotBots 07 Conference, pp. 1. USENIX Association, Berkeley, CA (2007)
5. Huang, S.-Y., Huang, Y.: Network Forensics Analysis Using Growing Hierarchical SOM. Research Center for Information Technology Information, Taipei (2013)
6. Geges, S.: Identifikasi Botnet Melalui Pemantauan Group Activity Pada DNS Traffic. Institut Teknologi Sepuluh Nopember (ITS), Surabaya (2013)
7. Chnadran, R.: Network forensics. In: Know Your Enemy Learning about Security Threats, 2nd edn., pp. 281–325. Addison Wesley Professional, Boston (2004)
8. Kaushik, A.K., Pilli, E.S., Josh, R.C.: Network Forensics System for Port Scanning Attack, Department of Electronics and Computer Engineering, Indian Institute of Technology Roorkee, Roorkee, India (2010)
9. https://www.wireshark.org/lists/wireshark-users/200910/msg00149.html. Accessed 20 July 2016
10. http://www.garykessler.net/library/file_sigs.html. Accessed 21 July 2016

Author Index

K.J. Kim et al. (eds.), *IT Convergence and Security 2017*,
Lecture Notes in Electrical Engineering 450,
DOI 10.1007/978-981-10-6454-8

Printed in the United States
By Bookmasters